decision theory

decision theory

d.j. white

Routledge
Taylor & Francis Group

LONDON AND NEW YORK

First published 1969 by Transaction Publishers

Published 2017 by Routledge
2 Park Square, Milton Park, Abingdon, Oxon OX14 4RN
711 Third Avenue, New York, NY 10017, USA

Routledge is an imprint of the Taylor & Francis Group, an informa business

Library of Congress Catalog Number: 2006040579

Library of Congress Cataloging-in-Publication Data

White, D. J. (Douglas John)
    Decision theory / D. J. White.—lst paperback printing
        p. cm.
    Originally published: Chicago: Aldine Pub. Co., [1969].
    Includes bibliographical references and index.
    ISBN 0-202-30898-7 (alk. paper)
        1. Decision making. I. Title.

T57.95W47   2006
003'.56—dc22   2006040579

ISBN 13: 978-0-202-30898-2 (pbk)

Contents

Preface

This report covers the main aspects of a research project, the objective of which was to ascertain the content of Decision Theory and its relevance to practical decision-making. Arising from the general nature of our research, which is concerned with the 'reasoning' content in selective action processes, is an interpretation of the word 'decision' which is not in accordance with popular usage. Reasoning requires premises and deductive steps, and this is identical with mathematical decidability.

It is not presumed to be the case that the ultimate aim of Decision Theory is to enable all actions to be decided. There will always remain elements of an undecidable nature in selective action processes. The boundary separating the decidable actions from the undecidable actions is an unknown quantity.

This report explores the various concepts, which have been brought to bear on the problem of decision, in terms of which action problems can be discussed, and the report explores the validity and use of these concepts for reasoning out actions.

We hesitate to use popular slogans such as 'improved decision-making'. This concept is examined directly and indirectly throughout the report. An attempt is made to give some meaning to this, but the report is by no means complete in this task.

We have used the term 'selective action'. This term refers to situations in which a selection mechanism takes some action (mechanism 'output') when presented with information (mechanism 'input') of a variety of kinds. Throughout this report the input contains a specified set of alternative actions but, in general, a mechanism may act, in the face of an input, without 'recognizing' the alternatives available. It will be seen, therefore, that Decision Theory is part of a more general Theory of Selective Action.

No attention has been paid to the 'mental processes' and other behaviour, which go on when an action is being selected. No attention has been given to the 'subconscious' or, at least, to mental processes which may be said to resolve problems without being identifiable. We all know of the expert who can act 'appropriately' without being able to retrace his thought process. This phenomenon cannot be ignored.

Formal reasoning, whether orally, by hand, or by computer, can be cumbersome. It is expensive in time, effort and money, and it is by no means certain that a 'well reasoned' argument should always replace the processes mentioned above. This applies, in particular, to the popular term 'quantification', and it must not be supposed that quantification, at any degree of refinement, is necessarily to be sought after, however meaningful in the context. We shall carefully examine the quantification concept. This is a point which has yet to be investigated in a scientific manner.

Arising from the complexity of the systems involved, certain parts will have to be treated as 'invariant black boxes' of which we neither know how they work nor intend to change their workings. Typical examples arise when 'values' and 'uncertainty' characteristics come to the fore.

Thus, from Savage's Theory of Subjective Probability, we can derive usable probabilities, but we need not inquire about the manner in which the input information is converted to such probabilities. Also using von Neumann's Theory of Value, we may measure the value of customer satisfaction, without determining how this value was, mentally, arrived at. It is, of course, a legitimate question to query the possibility of erroneous information processing, but this is a question on which we only touch lightly later on.

In each of these cases, the results may be used to change a person's selection in a given situation. It may, therefore, be thought that, contrary to the passive input–output analyses of the natural sciences, the results of Decision Theory will influence the theory itself. This is, however, not so, because it depends specifically on a careful identification of the input information. Thus, if we make a person aware of certain probabilities and values of specific elements of a complex problem, then he may very well select a different action to the one he would otherwise have chosen. This, in itself, does not make the changed selection any better than the original one. The principle we need to invoke to establish this is that the characteristics of micro component behaviour are to be accepted as reliable and that characteristics based on choice in a complex problem are less reliable. As an example, consider an investment expert. He may be able to give probabilities to the outcomes of individual shares, but he may be very poor in giving probabilities to the outcomes of combinations of these investments. The information content is effectively the same initially, but the processing differs. The theory is thus a theory of invariant behaviour, centring on the isolatable components of the decision problem faced, and dependent on the input information specified. This is an important point because we cannot compare alternative actions selected under different information inputs.

Perhaps some of the points made in the report could have been a little more clearly with the aid of well constructed examples. This is a task for the future: time limitations did not allow us to include many examples in the text.

Perhaps we can close with the following important remark. The research is concerned with the development of a language and its structure for the purpose of action analysis. The analyst requires formal unambiguous premises before he can attempt to reason out an action. The businessman very seldom formally exhibits his premises or his deductive steps. Perhaps this report will shed some light both on the difficulties of both and help to bridge what, sometimes, seems to be a wide gulf between the two.

Many points referred to in the report have arisen in discussion with many people encountered during the period of the research and I would like to thank them for their valuable time. In particular I would like to thank John Norman for his sympathetic ear in listening to my meanderings for these two years and for his appropriate criticisms when required. I would also like to thank Professor Johnston of the University of Manchester, without whose support and encouragement the project would not have been possible.

D. J. WHITE

CHAPTER 1

Some Concepts and their Interpretation

It has become fairly evident, as a result of work carried out in the general areas of decision, both from assessments of published work and from contact with actual decision-makers (and here I used the word 'decision-maker' in its commonly accepted sense), that, contrary to certain scepticism held by certain people, it is absolutely essential, by the very nature of the work and by pragmatic aspects, to consider the semantical side of our subject. This we shall do throughout, in relationship to uncertainty, value, etc., but in this section we shall be primarily occupied with the concept of 'decision'. As a result of the analysis it will be necessary to consider carefully the retention of the term 'decision-maker' as generally used. If it seems inadvisable to modify its usage, then we shall term our particular usage of decision 'logical' or 'mathematical' decision.

An essential pre-requisite of an occurrence of decision is the existence of a motivating state of ambiguity. Such a state may contain propositions whose truth is known and some whose truth is unknown. We may have 'partial'[1] ambiguity, although one term 'ambiguity' will suffice. Smith (64) takes the view that decision is simply the resolution of ambiguity, although the resolution of ambiguity may be achieved without deciding, in any logical sense. Thus in a problem $Q = \begin{bmatrix} a \\ b \end{bmatrix}$, the ambiguity can be stated in the form 'which act shall we take'? The choice of, say, a, is said to resolve this ambiguity. We may only wish to resolve some of the ambiguities. Thus, if the truth values of certain informational propositions are not known, we may still wish to select an investment opportunity without removing the informational ambiguities! We may not know the operating cost and yet still choose the investment.

Should the word 'ambiguity' be ambiguous, it is worth noting that the usual usage, in the sense of 'ambiguity of meaning of a

[1] The proposition '$x = \alpha, y \geqslant \beta$' is partially ambiguous.

1

proposition', is consistent with 'selection from a set of alternatives' (our usage) if a particular meaning corresponds to one alternative. Thus if we say '$c(a) = \cdot5$', c may refer to a cost calculated by any of several methods, and unless we know the method the proposition is ambiguous. Giving a meaning to the proposition is equivalent to asserting that a certain method was used in calculating $c(a)$.

The above paragraph has much more than an academic content for, later on, we shall see that in analyzing our primary decisions (e.g. whether to accept a certain investment proposition) we have to make secondary decisions, which are equivalent to selecting certain statements as being true. The truth or falsity of these statements can have a significant bearing on the appropriateness of any decision based on them and hence their analysis is just as important as the analysis at the primary decision.

One noticeable occurrence is the use of 'decision' as being synonymous with 'action' or 'alternative'. The use of 'the set of alternative decisions' is quite common, when 'the set of alternative actions' is meant. This may be of little significance in practice, but from our point of view we can afford no ambiguity of expression. The set of alternative actions identifies the ambiguity and the resolution of this ambiguity constitutes the decision process and culminates in the decision.

Another noticeable occurrence is the use of the term 'decision' when the term 'choice' is meant. Let us quote Dunlop (26), viz:

'There can be choice without decision; but there cannot be decision without choice.

'Throughout economics, that miserable puppet, the economic man, is allowed to choose but seldom if ever to decide. His choice, or choosing, is determined by his 'preference', just as choice of a strategy (not always clearly, in use, distinguished from move) is like collusion determined by rules, frequency ratio probabilities and self interest in the "Theory of Games". Colloquially, and more trivially, we encounter "take your choice", "make your choice", in each of which "decision" and an indefinite article can be substituted. If auxiliaries are to be necessary, we shall see clearly that for both decision and proof, the appropriate infinitive is "to arrive".'

These statements are extremely relevant to a theory of decision, and the key to our problem lies in the very last phrase, viz: 'to arrive'. Logical decision requires a linking of the state of ambiguity to the act of selection by a set of identifiable unambiguous cognitive operations. Certainly the operations must be unambiguous themselves, otherwise we cannot carry out the process of arriving in a deductive sense. As we shall see later, the problem of 'identifiability' is severe and raises questions not only of the possibility of identifying operations but also of whether they are cognitive or not. Logically speaking, unless they are cognitive, then the decision-maker can hardly be said to be deciding. It is to be expected in practice that resolution of specific ambiguities will be achieved by a composite mixture of decision and choice, and, in fact, this has become obvious from encounters with decision-makers.

A problem situation begins with the recognition of a primary ambiguity, e.g. 'shall we add to our production capacity?' In resolving this problem, secondary problems are generated, e.g. 'what information shall we get?', 'how thorough should we be in our analysis?' Quite obviously, over a time period, between the recognition of the primary problem and its final resolution, we can get a very complex mixture of secondary, and lower order ambiguities. The resolution of a problem may, therefore, involve a composite mixture of selection processes, some of which are cognitive and identifiable (which we term 'pure decision') and some of which are not so (which we term 'pure choice'). In general, the gap between the primary ambiguity and the primary selection will be covered by a set of selection processes, some being pure decision and some pure choice.

Provisionally, we say that ambiguity Q, in a state of knowledge K, is resolved by a pure decision process if there is an identifiable unambiguous cognitive operation θ such that

$$(Q, K) \xrightarrow{\theta} q \, \varepsilon \, Q. \tag{1}$$

θ has to be such that q follows deductively from (Q, K) via θ. It is not enough, for example, to say that 'if act a is represented by $[a_1, a_2, \ldots a_m]$ then θ is identical with forming the expression $\Sigma \alpha_i a_i$'. θ has to be completed by exhibiting that a formal rule exists, such as 'choose the act with the largest value of this expression'. The first form of θ establishes that $\Sigma \alpha_i a_i$ is a sufficient index for choice, but it does not follow that the second form holds. Given, say a problem

$Q \equiv \begin{bmatrix} a^1 \\ a^2 \\ a^3 \end{bmatrix}^1$, he may choose that act with the median value of $\Sigma \alpha_i a_i^j$.

[1] a^1, a^2, a^3 are alternatives.

3

There is nothing to prevent θ generating a sequence of ambiguities $Q_1, Q_2, Q_3 \ldots$ as pointed out previously. Thus if a is 'location and size' of warehouse, θ may required that operating costs (a_1^j, capital costs (a_2^j) and service factors (a_3^j) be evaluated first of all (thus posing a problem Q_1, the solution of which gives a^1, a^2, a^3) and then that the final choice be made on the basis of the minimum value of $\Sigma_i \alpha_i a_i^j$. θ can have many forms depending on how Q is initially presented. If θ is ambiguous or unidentifiable at any stage then the process is only partially decision. A better term for the common decision-maker is partial decision-maker, although this is not likely to gain acceptance in usage.

The selection of an action is not necessarily followed by its implementation. If, at the time of selection, any cognitive doubts arise as to whether it will be implemented, then it is not correct to say that the act has been selected. Good (30) introduces the concept of 'the amount of deciding (although "selecting" would appear to be more appropriate) in mental event E in favour of act F', where the probability of implementing the act serves the purpose of quantifying the amount of deciding. This has a direct bearing on probabilistic choice mentioned later, but this use of 'deciding' is not the same as ours and is more like 'a measure of mental commitment to act F'. However, this may not tell the whole of the story, for a person may be absolutely convinced that he will implement a certain action but be prevented from doing so by unforeseen circumstances. If, between the selection and the implementation, he has cause to reconsider this selection, it simply triggers off another selection process.

Common language, in discussions of problems, makes use of the concepts of 'preference' and 'choice'. So-called economic man is said to have 'preferences' for 'bundles' of commodities. A bundle may be denoted by a vector,

$$q = [q_1, q_2, \ldots q_n]. \tag{2}$$

Such preferences are not necessarily decidable. Although a preference function describing choice may be derived, using input–output analysis techniques, the decision-maker does not necessarily know this function. As we shall see later on, the foundations of a pragmatic theory of decision have to centre on certain basic types of preference characteristics. Although these preferences may not be decidable, we can use them, and, indeed, recommend their usage, in the analysis of complex problems, of which they become component characterizations. We can use them to introduce decidability into complex problematic situations, although they may not, themselves, be decidable. If he does cognitively use a preference function, such as $\Sigma \alpha_i q_i$, then we would ask how he derived this and whether or not it was decidable.

There are many instances of the use of linear decision functions in priority scheduling of production, and these decision functions are usually decidable by finding that one which optimizes production performance. There are, however, situations in which such rules are chosen, subjectively, by giving weights to each factor because of the subjective importance of each factor, e.g. such a case arises in giving priority to people on council house lists, or even to the selection of personnel.

If such a preference function is constructed by observing his choice behaviour, then, although this can now be used to decide his future choices, it does not mean that such a function was actually being used, cognitively, prior to this. This is certainly likely to be the case in personnel selection problems.

In order to illustrate the use of preference characteristics (cognitive or otherwise) in introducing decidability into problem resolution, let us consider a simple case, in which the commodities, $[q_i]$ are to be given the interpretation of physical goods, and let the decision-maker be given a quantity of money, x, and told to spend the lot on these commodities. Once the preference order for the q's is known, the actual selection can be 'decided'. In fact, if a value function $v(q)$ has been determined, the selection is decided by solving the following problem, in which p_i is the price per unit of commodity i:

$$\text{maximize } v(q) \text{ subject to } \sum_i q_i p_i = x, \quad q_i \geqslant 0. \qquad (3)$$

This shows that, although his preferences for the q's may not be decidable such preference behaviour can be used in decision analyses, by using the preference functions as premises for decision in a different class of problem. This is an important point, for it is not meant to be implied, by 'pure choice' that the results are not usable in deductive propositions, only that such choice may not, itself, be deduced.

If, however, he is given the money, x, and not instructed to spend the lot, then, for some reason or other, he may treat money as a commodity. The commodity vector is then:

$$q = [q_1, q_2, \ldots q_n, z].$$ (4)

The selection can be then decided by solving the following problem, where $v(q)$ is again inferred from observation of choice in terms of q:

$$\text{maximize } v(q) \text{ subject to } \sum_i q_i p_i + z = x, \quad q_i \geqslant 0.$$ (5)

A common difficulty occurring throughout the field of practical decision-making is the problem of determination of consequences, and this is discussed later in the report. It is, however, important to realize that it is closely tied up with the preference aspect, since, when someone is presented with, say, a commodity vector, parts of it may not, in themselves, constitute an ultimate consequence and the ultimate consequences may or may not be decidable. Whereas the problem (3) may be treated as indecomposable, problem (5) may be decomposable, in that some consequences of q may be partially decidable. Thus the future consequences of z, being a component of q, may be deducible in terms of some future q's. We have, in effect, interposed an extra identifiable unambiguous operation between the problem and the selection. There is nothing to prevent the decision-maker from putting some preference ordering on $\{q\}$, but this does not mean that he is informed about the consequences of z. If such information is decidable, then this may convert his choice activity into a partially decidable activity. It may be said that the possibility of converting choice into partial or full decision is the aim of decision theory.

Choice is the end point of a selection process, and preference results in choice. Choice is observable (at least in most cases) but preference has to be inferred somehow. We shall return to this later in the context of theories of choice.[1] Churchman (20), says that we cannot necessarily infer from observing a person's choice that it is a 'preferred' choice, although it is extremely difficult to produce a satisfactory operational definition of 'preference'. Certainly there are situations in which a choice definitely does not require preference. Thus if we consider randomized selection, in which there is a specific probability (achieved operationally by some well defined mechanism such as coin tossing) that a certain alternative will be chosen, we get 'choice', but would not infer 'preference'. We could extend the

---

[1] 'Choice' and 'selection' are interchangeable words. The distinction sometimes made is that humans choose or select, but mechanical devices can only select.

concept so that, if 'prob (a)' means 'a is chosen with probability $p(a)$',

$$(\text{prob } (a) > \text{prob } (b)) \rightarrow (a \text{ is 'preferred' to } b). \qquad (6)$$

We shall discuss the difficulties, arising in interpreting 'preference', more fully later on. At this stage it is useful to note that this is only one example of the difficulties in giving meaning to various concepts, and it is important to look at these problems in general. Another typical example arises in the attempt to give a meaning to 'probability'. Invalid answers to these problems can provide a severe stumbling block to the development of a pragmatic theory of decision.

Churchman (20) says:

'the meaning assigned to a concept is a decision assigned to serve certain interests.'

If he means 'decision' and not 'choice', then interests will determine the meaning. This sense is probably not implied by Churchman, and, although the meaning is motivated by the interests, it is not determined. James (39) echoes the same point of view in stressing the need to consider the 'cash value' of a concept, being typical of the pragmatic line of thought, although, again, deciding the cash value may be impossible.

Logical positivists (see Dineen (25)) take the view that meaning is identical with verifiability of the truth of a proposition and hence identifiable with the operations which carry out the verification. Thus 'a is preferred to b' may mean nothing more than 'a was chosen when the choice of a or b was offered'. It is important to note that, if the verification operations do result in verifiability of truth or falsity, then we are, in effect, choosing the meaning of 'preference' by choosing the verifying operations. It is not, therefore, question of giving meaning to 'preference', but solely a question of giving a label to an operational construct in the context of choice, the only prerequisite being that we do not stray too far from popular convention. It is rather as if we have some idea of what 'preference' is and wish to make it more precise.

If K is the state of knowledge, and Q is the ambiguity to be resolved, then, because preference is not strictly confined to dichotomous ambiguities (see Savage's Minimax-Regret criterion), we ought to write

$$qP(Q - q) \text{ rel. } K, \tag{7}^1$$

which is to be interpreted as 'q is preferred to the residual alternatives in Q relative to the body of knowledge K'. We call this 'relative preference'.

Another way of giving meaning to preference is to identify the operation θ which converts Q, given K, into the ultimate choice q. This is still consistent with the Logical Positivists' dogma.

We shall discuss the concept of 'randomized selection' later on, but a few words are perhaps in order here, arising from our consideration of probabilistic choice. We have already said that an issue may be partially decidable in that certain 'pure choice' stages will intervene between the recognition of a problem and the final resolution of this problem. However, the use of randomized, identifiable, selection processes involves a multivalued logic because the choice is not 'deducible', in the mathematical sense, from the initial premises. On the other hand it is weakly deducible in that, if we allow for an extension of the usual dichotomous truth values, $T = 0$ and $T = 1$, to cover the range $0 \leqslant T \leqslant 1$, we can, using probability calculus, deduce corresponding truth values for any propositions stemming from the propositions 'the truth value of "a will be selected" is $T(a)$'. In short, therefore, probabilistic selection can give rise to weak form of decision since it can be used within a language of probabilistic (or weak) decidability. Thus if act a results in a probability density function, $f(a, r)$, of returns r, then we can calculate, for a given problem Q, the truth value of the proposition '$r = r_0$'; in a two-person game we can calculate the truth values of returns for each act one player takes, given the randomized choice behaviour of the other player; in a multistage process in which the outcome of one act is another problem, the choice in relation to which is probabilistic, similar truth values can be deduced.

The occurrence of partial decision in practice is very self evident. The simplest form exists when Q is reducable to $Q' \subset Q$, i.e. it can be deduced that the alternative to be selected is in a smaller area of ambiguity Q', contained in Q. Wald (76) used this in his concept of admissible sets, in which no alternative can be deduced to be preferable to any other, and within which there exists, for each alternative not in the set, an alternative at least as good.

Markowitz (47) uses the same approach in his 'efficient set'

---

[1] c.f. equation (19) Chapter Two.

approach. Beyond such deductions it is then purely a matter of choice. A slightly stronger form would attribute certain likelihoods to each $q \, \varepsilon \, Q'$ as to its value or to it being chosen.

As an example, if x is chosen rather than y if $c(x) < c(y)$, and if $c(x) = c(y) < c(z)$, then this criterion reduces

$$Q \equiv \begin{bmatrix} x \\ y \\ z \end{bmatrix} \quad \text{to} \quad Q' \equiv \begin{bmatrix} x \\ y \end{bmatrix} \subset Q.$$

A further criterion is needed to resolve Q'.

There appear to exist, in practice, two different types of problematic situations. In one the alternatives are given and the decision-maker is required to make a suitable choice. In the other, part of the problem involves the search for alternatives. Ackoff (1) refers to the types as 'evaluative' and 'developmental' respectively.

On page 1 we have defined a state of ambiguity to be one in which a well defined set of alternatives exists. Bullinger (15) says that a person is never sure what he is deciding. The difficulties arise because of the failure to recognize the existence of several sorts of selection problem intervening between the primary ambiguity and any primary selection. For example, if we ask ourselves 'what is the problem?' this may be considered by Bullinger to illustrate his proposition. Further reflection however, leads us to realize that this is therefore our initial ambiguity and not the selection within the problems. We might be even more basic and ask ourselves 'do we have a problem?' and this, then, constitutes our initial ambiguity. It may be that Bullinger is concerned with some form of activity generation, which may be incognitive and for which the use of the term 'choice' is inadmissible since 'choice' requires a well defined state of ambiguity.

Whatever the answer may be, logical decision does require the existence of a well defined state of ambiguity and hence a well defined set of alternatives.

It is possible to query the problem of definition both of the 'set' of alternatives and of the form of 'an' alternative.

In the first example quoted, the set of primary alternatives would be identical with the set of all possible problems. Whether we can conceive of them is entirely outside the issue; the point is they are defined implicitly. They will inevitably be reduced, by a process of selection, to a small set, if defined extensionally (such as build a factory or put in an O.R. department or raise staff salaries), or to a large set, if defined intensionally, such as choice of product mix

$$x_1, x_2, x_3, \ldots x_n \quad \text{where } \sum \alpha_j x_j \leqslant \beta_i, i = 1, 2 \ldots m.$$

One might very well query the purpose of recognizing these ambiguity phases in a primary selection process. We have said that the purpose of decision analysis is to see how much selective behaviour can be made decidable. It, therefore, requires a complete decomposition of an activity into its various components, including the identification of each ambiguity and its form. Only then can we carry out any analysis. Viewed from this point of view the so called nebulous selection processes, when decomposed, become a little more comprehensible.

A central concept in decision theory is that of 'measurability'. 'Measurability' cannot be divorced from 'predictability', since the only purpose of measuring is to be able to predict certain events. If such measures are ambiguous in their representation, either the ambiguity has to be treated as such, or it has to be removed. Otherwise we have no premises for any analytical reasoning. A general occurrence of this sort of thing arises in the continual battle of 'ordinality' versus 'cardinality'. Ranking of a set of realizations of a particular factor allows only partial decidability, whereas cardinality, if feasible, removes much ambiguity and allows a higher degree of decidability. Thus if one person is better than another in every relevant respect, we may deduce that he is the appropriate choice for a post; if, however, he is better in some respects, and worse in others, we cannot deduce he is the appropriate choice, unless we use a more refined measure.

Admittedly the ability to measure the various aspects and to produce a decision function poses a problem in itself. However, at this stage, our point is simply that ordinal measures permit a very limited form of decidability. As we shall see later, they have another form of ambiguity, since ordinal measures are, by necessity, relative to a specific set, and hence ambiguous in their use outside these sets (the rank of an item changes with its reference set).

So far we have concerned ourselves with the purely logical notions of decision, partial decision and choice. Since the purpose of decision analysis inevitably impinges on the physical and psychological aspects of selective processes in general, particularly if we are going to suggest changes in such behaviour, we ought to consider such aspects. Dunlop (26) says that there are, to the extent of present knowledge, four phases in decision (although I am not so sure this relates only to logical decision). These are hesitancy, knowledge, striving and willing. Presumably 'hesitancy' signals that a problem, or ambiguous state exists; 'knowledge' would contain information on the specification of the ambiguity as well as information relevant to the ambiguity resolution; 'striving' would correspond to an attempt to decide between the alternatives; 'willing' would correspond to a will to use any alternative chosen. 'Striving' would have to include a decision process itself, since it would involve informational activities.

It is one thing to break such selection processes up into a set of independent phases, and entirely another to be able to determine precisely what goes on in these phases. We shall meet this difficulty in the theory of choice, in which, for example, determining the knowledge content presents severe difficulties. Nevertheless, consideration of cognitive aspects of selection will be important in any satisfactory pragmatic theory of decision: it may be that we want to simulate selection processes mechanically, and hence remove mental effort, bearing in mind risks that the analogues may not be perfect; it may be that we want to convert a pure choice situation into a full or partial decision situation and hence add to mental effort. Such considerations will, in conjunction with the consequences of any such change, be important in any definition of 'improved selection processes'.

It may be that mental effort considerations are negligible at the higher level problem area. There will, however, be certain areas of selection activities in which the answer is not so obvious. We would not, presumably, try to add much analytic content to the process of choosing whether to write a letter now or ten minutes later. There are great difficulties of getting to grips with this sort of problem, and these must, for the purely logical purpose of this report, be omitted here. Almost certainly not a great deal of work has been done in this area, even by psychologists. One work by a statistician (Good (30)) is concerned partly with 'decisionary effort in deciding in favour of an act F in a mental event E', and this may merit further consideration, although our usage of 'decision' is not Good's usage.

As a corollary of the above remarks it is likely to be singularly difficult to determine whether a selection process is a decision process, or to what extent decision does occur. This is magnified by the possibility of 'selections within selections'.

The above section defines what we mean by decision in the logical sense and discusses some of the relevant concepts and their meanings. Common usage in technical discourses is sometimes not consistent. Nevertheless we are not likely to be able to proceed far in the analytic treatment of problem solution without restricting ourselves to a logical concept of decision. We shall now proceed to give a coverage of some of the theories relating to choice, which form the basis of most research work in this area. These theories relate to choice, value and uncertainty. As will be seen later, these are not independent quantities, from one point of view at least, depending on whether we take the so called subjective or objective points of view. It will be seen that there is no necessary connection with decision in any of the theories. The theories do not imply that apparent criteria are actually cognitive criteria. They do, however, form the foundations of the procedures by which decidability can be introduced into problem solving.

Summary

In the introductory chapter we give consideration to certain concepts important in the development of Decision Theory, viz.: ambiguity, decision, action, choice, primary problem, secondary problem, problem, mental commitment, knowledge, decision operation, preference, consequences, partial decision, probabilistic decision, relative preference, decidability, weak decidability, definition of problem, hesitancy, striving, willing, decisionary effort.

Decision-making is concerned with resolution of ambiguity and is synonymous with mathematical decidability. Ambiguity arises when a question is posed. Such a question may be concerned with the meaning of a proposition. Selecting such meanings is no less a matter of choice than the problem for which the answer is needed.

Popular usage of 'decision' and 'action' (alternative) is not correct and this point is clarified.

The major distinction drawn in this work is between 'decision' and 'choice'. Choosing need not entail deciding. Decision does not exist unless the process of converting the primary problem to the final choice is identified. We may be able to provide a behavioural description of this, but it does not follow that the subject is deciding. Nonetheless behavioural characteristics can be used to add decidability in extended problems.

The resolution of the primary problem involves the generation of secondary problems (e.g. choice of information or model). Such a complexity is resolved by a combination of decision and pure choice.

A formal presentation of our main theme is in terms of the manner (decision operation θ) in which a problem, Q, is converted, in the context of a knowledge content K, to the final choice q. θ may be a problem solving operation generating secondary problems.

There is a sense in which a person may choose an action but not be committed to it. We do not concern ourselves with this but the theory can be extended to cater for this.

The concept of 'preference' is a central one in Decision Theory. This is discussed fully later on, but we observe here that preference may be apparent and not actual. Such characteristics may themselves be decidable, but, even if not, they can be used to add decidability. We illustrate this with the usual economist's 'commodity bundle'.

The concept of 'consequences' also plays an important role in Decision Theory. The failure to pursue consequences far enough adds 'partial decidability' to our decision-making.

The third form of decidability is 'probabilistic decidability' (or weak decidability) and 'preference' becomes meaningless, in this case, as a means of describing choice behaviour. We are concerned so many times with the meaning of certain variables (probability, value, information, etc.), that we need to examine the ways in which meaning is given to a concept. The operational, predictive, and pragmatic views are discussed. Consistent with our definition of decision, the giving of a meaning to a concept is a decision itself. There have been attempts to put cash values to concepts. We illustrate our points in the context of 'preference'.

The idea of 'relative preference' is discussed since this has an important bearing on the notion of optimization (see Chapter 2, Savage Minimax Regret Criterion). There are two ways in which primary problems can be generated: (a) the alternatives may be given, (b) desires may be given and alternatives searched for. Our starting point differs, but, in each case, decidability requires an explicit statement of some problem. Thus in case (b) the primary problem still needs to be stated, although its definition may be weak.

The reasons for paying so much attention to the secondary problems are that they have an influence on the resolution of the primary problem and that the process of identifying them has a bearing on the decidability of the primary problem.

The problem of definition goes much more deep than we have discussed in the context of problem definition. It is related to ambiguity, and we discuss, in the context of defining measures, the ambiguities and their effects on decidability, which arise from ordinal and cardinal measures.

Finally we give brief consideration to the physical and psychological phases of the decision process. The four phases so far identified are: hesitancy, striving, knowledge, willing. The cognitive aspects are important in Decision Theory. The 'mental effort' should also be considered, since one way of improving decision-making lies in reducing mental effort. The concept of 'decisionary effort' is briefly discussed.

CHAPTER 2

Theories of Choice, Value and Uncertainty

We shall be concerned, in this chapter, with some of the theories centring on the concepts of choice, value and uncertainty. It will be seen that some theories are 'closed' in that they are concerned with all three concepts, at the same time, as interdependent concepts. Most theories assume theories of at least one of the concepts. Thus Savage's general theory considers choice, value and uncertainty as part of one integrated theory to be investigated; von Neumann assumes an objective probabilistic representation of uncertainty; Milnor assumes a value theory for consequences. We shall see later that even the so-called isolated objective theories of uncertainty require some inference about choice behaviour. As a consequence of this there must be a high degree of interdependence, in principle, among the above three concepts.

2.1. Choice and Order Relations

We have already said, in Chapter 1, that 'choice' has no necessary connection with 'decision', in that all choice is not decidable. In order to construct a pragmatic theory of decision we do, however, need a starting point. Such a starting point is the observation of choice in specific problematic situations. The development of the mathematical theory of value is accomplished by introducing the idea of 'order relations'.

On a purely formal plane, choice is represented by one of the relations $>$, $\geqslant$ or $=$. '$A \geqslant B$' means 'A is chosen when A, B are presented for choice'. We may or may not wish to infer preference, but the person could be indifferent between A and B and yet choose A. If he states verbally that he prefers A to B then we say $A > B$. If he states verbally that he is indifferent between A and B, then we say $A = B$. We can infer neither of these simply from observing a single choice. It is possible to extend the choice over many repetitions of the pair A, B, and if A is chosen more often than B, we may wish to infer that A is preferred to B in that the person feels A is probably better than B.

Since every theory we shall consider must hinge essentially on observing choice behaviour in specific circumstances, this difficulty constitutes one of the most severe ones in decision theory in general. Obviously the inferences we draw from observing behaviour will only assume importance if they are to be used as premises for further selection behaviour. If we ask a person to choose between A and B and he chooses A, can we infer that he will always choose A rather than B? If he were deciding, probabilistically or otherwise, and we could determine his decision operations, we could answer this question. However, we are making no assumptions of this kind as yet. We have no grounds for drawing any such inferences unless we know more about the selection process itself. An attempt to repeat the experiments several times is beset with difficult inferential problems itself because of the inability to isolate the experiments, although, no doubt, evidence of a series of experiments may be more acceptable to the experimenter himself.

Whichever operational meaning we give to $A \geqslant B$, the operational meanings of $A > B$, $A = B$ are derivatives of the meaning of $A \geqslant B$. Thus, $A > B$ is equivalent to $A \geqslant B$, $B \ngeqslant A$, and $A = B$ is equivalent to $A \geqslant B$, $B \geqslant A$.

Let us now return to the purely formal aspects, and, in particular, to the use of 'order relations' in axiomatic theories of choice, value, and uncertainty. This is essential because many of these theories assume familiarity with the concept of order relations. Chipman (19) gives a comprehensive coverage of general relations and their properties. If R is a relation between elements of a set χ, the relation R^2 is to be interpreted as 'xR^2y if there exist a $z \, \varepsilon \, \chi$ such that xRz, zRy'. The relation R' is to be interpreted as '$xR'y$ if yRx'. The set Δ is the region of $X \, (\equiv \chi \times \chi)$ of the form (x, x), $x \, \varepsilon \, \chi$. $\tilde{R}$ is the complement of R in X. The following properties of R are considered.

(i)	Transitivity	$R^2 \subseteq R$
(ii)	Density	$R \subseteq R^2$
(iii)	Indempotency	$R^2 = R$
(iv)	Triviality	$X \subseteq R$
(v)	Vacuousness	$X \subseteq \tilde{R}$
(vi)	Comparability	$X \subseteq R \cup R'$

(vii)	Directedness	$X \subseteq R'R$
(viii)	Symmetry	$R' \subseteq R$
(ix)	Asymmetry	$R' \subseteq \widetilde{R}$
(x)	Non-Symmetry	$R' \neq R$
(xi)	Reflexivity	$\Delta \subseteq R$
(xii)	Irreflexivity	$\Delta \subseteq \widetilde{R}$
(xiii)	Antisymmetry	$R \cap R' \subseteq \Delta$
(xiv)	Contradirectedness	$X \subseteq RR'$

To these we add:

$$(\text{vi})^*: \ \widetilde{\Delta} \subseteq R \cup R'$$

Let us re-interpret some of these properties. (i) means that if xRy, yRz, then xRz; (vi) means that either xRy and/or yRx; (vii) means that, for all x, y, there is a z such that zRx and zRy. Let R be identical with the normal arithmetical inequality $\geqslant$. Then we have: (i) $x \geqslant y$ and $y \geqslant z$, then $x \geqslant z$; (vi) for all x, y either $x \geqslant y$ and/or $y \geqslant x$; (vii) for all x, y there is a z such that $x \geqslant z$, $y \geqslant z$.

Certain of these properties are essential to a logical development of decision theory, for on these depends the validity of any deductions we can make concerning, particularly, the validity of value and uncertainty measures.[1] Let us consider the transitivity and comparability axioms, since these are the most common ones discussed in the literature on value theory.

Let $A \geqslant B$ have the operational definition 'A chosen when A and B are presented for choice'. It is not generally accepted that, in practice, if a person chooses A rather than B, and B rather than C, he also chooses A rather than C. This may be for one of several reasons; (a) the 'qualifying conditions' at the time of choice may differ in each choice situation; (b) the mental selection process (whatever this is) may function differently in each choice situation; (c) he may actually be randomizing, and hence breaking the intransitivity condition with some degree of probability.

May (49) constructs an example in which group selection is derived, in an explicit manner, from individual selection. Each person is asked to give a strict preference ranking. For three people the individual

---

[1] In this section we consider only those properties particularly important to value and uncertainty theory. The other properties are, however, particularly pertinent to the general problem of measurability. Thus, for example, the cardinal measure is very much geared to properties (ii), (vii), (xiv) as well to properties (i) and (vi).

choices are assumed to be transitive, and the majority rule determines group paired selections. The individual situations may be described as follows:

Person	Ordering
1	$A > B > C$
2	$B > C > A$
3	$C > A > B$

The majority rule then gives, pairwise, $A > B$, $B > C$, $C > A$, and this is intransitive.

May gives another example in which intransitivity arises because of variations in the basis of comparison from one choice situation to another. He quotes the case of a pilot faced with three choice situations, the choices being those of flames or red-hot metal, red-hot metal or falling, falling or flames. It was found that his preferences were flames rather than red-hot metal, red-hot metal rather than falling, falling rather than flames. Thus his behaviour was intransitive. Further analysis uncovered the facts that his preferences were, respectively, based on the factors: heat, support, probability of death. An analysis of business behaviour would almost certainly uncover similar situations, particularly when comparisons of a heterogeneous collection of investment opportunities are involved: thus when choosing between three candidates for a job, A may be preferred to B because he has better qualifications; B may be preferred to C because he is better dressed; C may be preferred to A because he is more industrious. It all depends on the more striking comparisons between pairs which may attract focal attention.

Although the transitivity axiom is the most criticized axiom, the comparability axiom is also queried. It is one thing to observe that a person chooses A rather than B, and it is another to say that any comparison is made. It all depends, as before, on what inferences we wish to draw from choice. It is quite often the case that a business man will say 'I have no logical basis for comparing one alternative with another'. The operation of 'providing a basis' is one step in converting choice into decision. Thus, given two alternative investment proposals (such as replace or do not replace a piece of equipment), if he has a formal way of providing information on which final choice is based, this provides an element of decision.

It is essential to allow for only 'partial' orderings of our elements. If we state '$A > B$', then this is a premise to be used in all analyses. We are not, however, obliged to state that either '$A > B$', or '$B > A$', or '$A = B$'. In other words some elements will be comparable, whereas others will not. Thus, if an investment opportunity A is defined by a set of net incomes $r_1, r_2, \ldots r_n \ldots$, and if B is defined by another set $r_1^1, r_2^1 \ldots r_n^1 \ldots$, it may be reasonable, to say '$A > B$' if $r_s \geqslant r_s^1$ for all s. It may be not reasonable to say '$A > B$' or '$B > A$', or '$A = B$' if $r_s \geqslant r_s^1$ for $s \, \varepsilon \, \Omega_A$, $r_s < r_s^1$ for $s \, \varepsilon \, \Omega_B$. The decision-maker can assert that '$A > B$', whatever the $[r_s]$, $[r_s^1]$ may be, but this may not be consistent with choice based on a knowledge of other consequences of the investments A, B: if we posit that all income is invested at a fixed rate of interest, α, and if he also is required to choose between two alternative quantities of cash after ten years, say, then his expressed choice between A and B in the first place, may conflict with his expressed choice of total cash quantities derived from A and B. We have introduced an element of decidability into the second situation which may make incomparable alternatives ultimately comparable.

'Comparability' is not a trivial concept and is central in the theory of decision, whose purpose is primarily to introduce a consistent form of comparability between alternatives. However, we have to take a starting point somewhere for our basic preferences and the set χ with which we commence could conceivably contain incomparable elements. For this reason incomparability necessitates the inclusion of partial orderings. It is worth noting that the works of Wald and Markowitz centre on the ideas of partial orderings. 'Pareto Optimality', a common term in economics, also centres on partial orderings.

2.2 Value

2.2.1 Value as an Order Preserving Homomorphism

A great deal of the early mathematical work on value (or utility) theory was concerned simply with finding mappings of a set of ordered elements into the real line in such a manner that the mapping ('homomorphism') preserves the ordering on the real line. If χ is a set of elements, ordered by a weak[1] ordering relation R (one which satisfies properties (i), (vi)*, (xi), and (xiii) of Section 2.1), then the

---

[1] This is too strong and we can replace this more fruitfully by a total order (satisfying properties (i), (vi), (x)). This avoids property (xiii) and allows xRy, yRx without $x \equiv y$. When χ is commodity space we need to be able to have x indifferent to y without, in some cases, $x \equiv y$.

search was for a real valued function, v, on χ such that if $x, y \, \varepsilon \, \chi$, then

$$(v(x) \geq v(y)) \rightleftarrows (xRy). \qquad (1)$$

It is essential, at this point, to refer to the points, made previously, about preference inferences and operational meanings of R. From observing of a single choice event we cannot infer 'a is preferred to b' in the sense that in the same situation, given a, b as alternatives, a will always be chosen. The decision maker may not be sure whether he really prefers a to b and we may wish to consider a weaker interpretation of R.

As a means of overcoming this difficulty Debreu (24) and Luce (45), who is, however, concerned with more than order preserving homomorphisms, have introduced the concept of probabilistic choice, and a corresponding relation R is based on this. The 'strength' of a person's preference is measured by the proportion of times he chooses a rather than b. Debreu's axioms are as follows.

Axiom 1: S is a set; p is a function from $S \times S$ to $[0, 1]$ such that $p(a, b) + p(b, a) = 1$.
Axiom 2: $(p(a, b) \leq p(c, d)) \rightleftarrows (p(a, c) \leq p(b, d))$.
Axiom 3: If $p(b, a) \leq q \leq p(c, a)$ there is a $d \, \varepsilon \, S$ such that $p(d, a) = q$.

$p(a, b)$ has the physical interpretation that a is chosen, rather than b, a proportion $p(a, b)$ of times.

Debreu deduces that there exists a real valued function u, on S, which satisfies:

$$(p(a, b) \leq p(c, d)) \rightleftarrows (u(a) - (u(b)) \leq (u(c) - u(d)) \qquad (2)$$

The content of (2) is entirely different to that of (1). From (2) we see, by putting $c = b$, $a = d$, that

$$(p(a, b) \geq \tfrac{1}{2}) \rightarrow (p(b, a) \leq p(a, b)) \rightleftarrows (u(b) \leq (a)).$$

Hence 'aRb' is consistent with 'a chosen, rather than b, at least 50 per cent of the time'. Of course, (2) goes a lot further than (1) since it effectively allows comparison of strength of preferences.

The probabilistic choice behaviour may be removable by further investigation of the alternatives. We have, however, to begin somewhere with behavioural characteristics in some standardized situation, and hence the behaviour described may form the premises for our later analyses.

One could voice some objections against the axioms. Thus if

19

$p(a, b) = 0$, $p(c, d) = 0$, $p(a, c) = 1$ (cases of certainty) then axiom (2) implies $p(b, d) = 1$.[1] Therefore '$aP'b$, $cP'd$, aPc' implies 'bPd', and this is violated by $a = 2$, $b = 3$, $c = 1$, $d = 4$. However, the general idea is useful and it remains to construct a consistent system.

This theory, as with many others, confines itself to dichotomous choice situations. Thus, given a, b, c, it says nothing about the probability $p(a:b, c)$ that a will be chosen when a, b, c are presented for choice. This depends solely on the order in which they are presented in pairs. We cannot deduce $p(a:b, c) = p(a, b) p(b, c) + p(a, c) p(c, b)$ unless we specify operationally that (b, c) are compared first and then the outcome of (b, c) with a.

Ackoff defines partial, weak, and chain orderings to be those which satisfy properties ((xi), (xiii), (i), not (vi*), ((xi), (xiii), (i), (vi)), and ((xii), (ix), (i), (vi)*) respectively. The usual arithmetical relations '$=$', '$\geqslant$', '$>$' are respectively partial, weak and chain orders.[2]

The fact that the elements of a set can be ordered does not mean that a real order preserving homormorphism exists. Debreu considers the ordering of the two dimensional Euclidean plane, χ, (in which x is a two dimensional vector (a, b)) according to the rules:

$$((a, b) R (a^1, b^1)) \rightleftarrows ((a > a^1) \text{ or } (a = a^1 \text{ and } b \geqslant b^1)). \qquad (3)$$

He shows that, if we allow a, b, to take any real value, no order preserving homomorphism exists between χ and the real line.

Chipman, in an extremely mathematical and comprehensive analysis, shows that if a 'total'[3] ordering (satisfying properties (i), (vi), (x)) exists over a set χ, and if the 'axiom of choice' holds (viz: if we divide χ into sets of equivalent elements, then the derived set is 'well ordered', i.e. can be indexed by an ordinal number), then the most general form of order preserving homomorphism is a lexicographic ordering over some, possibly cardinal dimensional, Euclidean space. It is possible, of course, to have a vector, not with n

---

[1] 'xPy' is defined by 'xRy, $y\tilde{R}x$'. $xP'y \equiv yPx$. In the arithmetical system $R \equiv \geqslant$, $P \equiv >$, $P' \equiv <$.

[2] A weak order, R, is not sufficient to be able to differentiate between elements, and we need to be able to derive the chain order, P, given by 'xPy' $\equiv$ 'xRy, $y\tilde{R}x$'.

[3] A 'weak' order is a 'total' order but, for a 'weak order, an element of χ cannot be indifferent to any other element of χ. Note that weak orders are used when the existence of values is being considered. However, it is too strong when we need to consider spaces, χ, in which we can have xRy, yRx, but not $x \equiv y$. This arises when χ is n-dimensional commodity space.

A weak order, R, is obtained from a total order, T, by generating an equivalence set Ξ from χ by $e \equiv \{y:xTy, yTx\}$ for any given x. The induced relation on Ξ is weak.

components, but with a component corresponding to any real number.

The concept of 'lexicographic' valuations deserves further consideration, both because of the existence of lexicographic rules in day to day control of business, traffic, etc.; and because of the problem of whether lexicography is, in principle, valid in any sense.

A general representation of lexicographic choice is given by two sets of hierarchical order relations, $P_1, P_2, \ldots P_n, \ldots$ and $I_1, I_2, \ldots I_n, \ldots$. Choice then follows the pattern: a is chosen rather than b if, for some s,

$$\text{`} a I_k b, \quad k = 1, 2 \ldots s, \quad a P_{s+1} b \text{'} \qquad \text{or} \qquad \text{`} a P_1 b \text{'}. \qquad (4)$$

In talking to certain decision-makers, in the course of this research, apparent lexicographic behaviour was found. One company is a strong believer of 'promotion from within', and, apparently, when a position becomes vacant, if at least one employee is a candidate for the post then no external search is made (although prior selection processes probably determine the possibility of a person being a candidate). There is a very strong value on internal promotion. Next, when it comes to choosing between candidates within the firm, formal qualifications are determinants of promotion. Should there still be a tie after this, the best man for the job is considered. Finally, if there is still a tie, the one who is least likely to upset other activities is considered. We thus have four order relations P_1, P_2, P_3, P_4 with corresponding indifference relations I_1, I_2, I_3, I_4. Thus if a and b were equivalent with respect to being employees of the firm, equivalent with respect to qualifications, and a was the better man for this particular position, but his leaving his present post would cause some organizational difficulties, although b had an adequate deputy ready to step into his place, a would be promoted.

There are other examples. Thus priority scheduling is used at the production level, in which orders are fed onto machines according to priority customer rating first of all, then according to present delay in order, then according to size of order and so on.

It is questionable whether lexicographic behaviour is apparent or actual. It may be that the spectrum of choice situations which have been faced may be such that choice conforms to such rules, but that an enlargement of the spectrum may not conform to these rules. In the first example it may be that, so far, no external candidate has been that much more qualified, formally or by experience, that he could overcome the relatively high, but nonetheless by no means infinite, weighting given to 'promotion from within'. It may be, in the production problem, that no low priority customer has yet been ten years behind on getting his order. One must feel that, in principle at least, much of the apparent lexicography is not actual lexicography.[1] Quite obviously some of the above choice situations are decidable to some extent and the question of lexicography can be solved using this. Thus optimum production rules, derived from consideration of the production consequences, are unlikely to be lexicographic, although optimum lexicographic rules may still be useful.

A common situation in which apparent lexicography arises is in the putting of a constraint on a problem. This is an extremely important point in all problem solving. Consider the problem of determining an optimum operating policy for production subject only to a budget constraint. The problem might then reduce to

$$\text{maximize }_\pi[\theta(\pi)] \text{ subject to } c(\pi) \leqslant z, \tag{5}$$

where $\theta(\pi)$ is some measure of output and $c(\pi)$ is the cost, both depending on production policy π.

This means that we have a lexicographic ordering for policies π, given by

$$v(\pi) = \begin{bmatrix} \max\ [(z - c(\pi))/|z - c(\pi)|, 0] \\ \theta(\pi) \end{bmatrix} \tag{6}$$

In the first instance all policies with the same first component value would be equivalent. Those with value 1 would be preferred to those with value 0. Those with the same first component values would then be differentiated according to the value of $\theta(\pi)$.

The reason for the lexicography lies in the failure to determine the consequences of the constraint '$c(\pi) \leqslant z$', on the rest of the system. Thus the maintenance budget can vary as z varies, and this has consequences for the system; other investment budgets are also influenced by z; thus varying z not only influences $\theta(\pi)$ but also subtracts from, or adds to, system performance in other areas. One way of trying to put a value on constraints is by using the Lagrange parameter method of valuing constraints. This is a restricted form of valuation. The procedure would be to fix λ, and then to

---

[1] See page 28, where reference to an extension of Walds' maximin criterion could conceivably give lexicography in uncertain situations.

$$\text{maximize } _\pi[\theta(\pi)-\lambda c(\pi)]. \tag{7}$$

For each λ, $\pi(\lambda)$ is determined and then $\pi(\lambda)$ maximizes $\theta(\pi)$ subject to $c(\pi) = c(\pi(\lambda))$. Usually the optimum occurs when $c(\pi) = z$ (although for general problems this is not so). The parameter λ, at the optimal point, is said to be 'the value of violating the constraint by one unit'. However, the real interpretation is 'if we attach an (additive) value of λ per unit of constraint violation, then the optimal solution to the constrained and unconstrained (modified Lagrangian) problems are the same.' It says nothing about either the true value of constraint violation measured by its effect on the system nor about any subject values of the decision-maker which gave rise to the constraint in the first place.

In principle at least, real valued order preserving homomorphisms are not necessarily existent. However, it may be that it is simply an academic question. Let us accept the boundedness of all the variables which enter into any choice situation and let us assume that the range of each variable is divided by a finite set of discrimination points. Consider the lexicographically ordered n dimensional Euclidean plane, E_n, in which $r\delta(k)$, $r = 0, 1, 2 \ldots N(k)$, are the partition points of the kth component. Suppose we now choose real numbers $a(k)$, $k = 2, 3 \ldots n$, such that

$$a(k)\delta(k) > a(k-1)N(k-1)\delta(k-1). \tag{8}$$

Then the value function given below will be equivalent to the natural lexicographic ordering over E_n with higher order components having the higher values.

$$v(r_1, r_2 \ldots r_n) = \sum_{k=1}^{n} a(k)r_k\delta(k) \quad (a(1) = 1). \tag{9}$$

We ask ourselves 'is lexicography an important concept in practice?' Has it any value itself for formal analysis of problematic situations or is it of mathematical interest only? From a practical point of view the answer may lie with the above derivation of an equivalent real value function, the point being that our tools for measuring variables do not produce a continuum of numerical values, and essentially produce values at some fixed distance, $\delta(k)$, from each other[1]. From a pure computational point of view, the use of lexico-

---

[1] This limit on measurability operates away the difficulties arising in theoretical discussions on the existence of real values, but we will discuss these later on just the same.

graphic operations may be easier than handling equivalent real valued functions such as given by (9). When comparing two alternatives lexicographically, the choice can conceivably be made at the first level, not necessitating consideration of other components. In the equivalent form (9) we only make one comparison, but we have extra additive arithmetical operations, to say nothing of the fact that every r_i has to be determined. This latter point is important, for some of the components may be difficult to measure at all and if we can resolve the choice early on some of them need not be considered. Thus the delays in satisfying an order may be so striking that other gains or losses need not be considered, even though the true value function is not lexicographic.

It is to be expected that some problems will necessitate consideration of every level in the lexicographic ordering, and some may be resolved at the first level. The answer to the problem of which is the better may then lie in the realms of probability theory.

Let us briefly discuss mathematical aspects of existence of value functions. In the light of the previous remark on measurability, all variables are countable and this guarantees the existence of real valued functions in practice.[1]

Chipman and Debreu consider conditions sufficient for a real valued (order preserving) homomorphism to exist. The treatment is primarily topological, and depends on the ability to map a region (in χ) of the form $(z : xRz, zRy)$ (where x, y are fixed), onto a region on the real line of the form $u \leqslant v \leqslant \bar{u}$, where u, $\bar{u}$ are appropriately chosen. Although we can always get a one-one mapping of Euclidean n space into Euclidean 1 space, we cannot always do it so as to satisfy this pre-requisite. Debreu demonstrates this.

Let us see how we can determine real value functions. Suppose we have an n dimensional vector representation over the field of rationals; as has already been stated, every variable has a rational measure. We can then number these with integers, in such a way that it is possible to determine in advance what integer corresponds to each vector, although this may be quite complicated. Suppose now we begin with the first vector, and use the following procedure. Assume we have allocated real values (rational or otherwise) to the first k numbers. Using the deterministic procedure of allocating numbers to our vectors, we can determine which vectors correspond to the first $(k + 1)$ elements and hence we can allocate a real number

[1] Providing a weak order R exists.

to the $(k+1)^{th}$ element, unique only to the extent that it preserves the order of the mapping. If $v(1)$, $v(2)\ldots v(k)$ have been determined, and if the $(k+1)^{th}$ element is better than s elements, L, and worse than $(k-s)$ elements, B, we can allocate any real number to the $(k+1)^{th}$ element providing

$$\max_{i \in L}[v(i)] < v(k+1) < \min_{i \in B}[v(i)]. \tag{10}$$

We can always continue to do this since our set of elements to be valued is countable. In practice even this is not required, since, we can, without loss, assume they form a finite set.

As an example, suppose we consider Euclidean 2 space over the set of rationals, $0 \leqslant r \leqslant 1$. Each vector is then of the form $(p/u, q/u)$ (reduced to a common denominator). Numbering is then lexicographic, first by u, and then by p, q subject to $p+q \leqslant u$. If at any stage a vector is repeated, it is deleted, since it has already been given a number. We thus get

$$\begin{aligned}
&(0,0) \to 1\,;\, (1,0) \to \ 2\,;\, (0,1) \to \ 3\,; &(11)\\
&(\tfrac{1}{2},0) \to 4\,;\, (0,\tfrac{1}{2}) \to \ 5\,;\, (\tfrac{1}{2},\tfrac{1}{2}) \to \ 6\,;\\
&(\tfrac{1}{3},0) \to 7\,;\, (0,\tfrac{1}{3}) \to \ 8\,;\, (\tfrac{2}{3},0) \to \ 9\,;\, (\tfrac{1}{3},\tfrac{1}{3}) \to 10\,;\\
&\qquad (0,\tfrac{2}{3}) \to 11\,;\, (\tfrac{2}{3},\tfrac{1}{3}) \to 12\,;\\
&\qquad (\tfrac{1}{3},\tfrac{2}{3}) \to 13
\end{aligned}$$

Suppose now we wanted to determine a value for any particular vector. Thus suppose we wanted the value of $(\tfrac{1}{3}, \tfrac{2}{3})$. We would only have to determine which were the first twelve elements to be valued, assign them values (by the same iterative process), and put any value on $(\tfrac{1}{3}, \tfrac{2}{3})$ to maintain consistency of order. We can determine a value for any rational vector in a finite number of steps, without having to consider all vectors. This is important, since we could not possibly consider all vectors.

It is quite obvious that so far we have achieved nothing of practical usefulness. There is certainly no uniqueness of such value scales, and hence no absolute meaning. We have simply indexed a set of elements so that the order of the elements is preserved by the index. What value has any such operation? As it stands it has no value. Anyone presented with two alternatives could not determine how the other person would choose without knowing effectively the person's choice behaviour, since to know that '$v(a) > v(b)$' would be no different to knowing 'a' would be chosen rather than 'b'. A listing of all possible alternatives would be needed, if all possible choices were to be predictable. Giving a table of values would not be, in principle, any different to giving a full list of actual choices. Any real advantage, from the predictor's point of view, would only arise if he were able to 'decide', from a smaller amount of information, how the person would choose. The 'forecaster' may be a computer taking over the role of the person in question.

The ability to compute values from a small number of premises is a central point in our theory and we shall deal with it, and relevant theories, in the next three sections.

2.2.2 Value Structures

We cannot discuss all the axioms and points of issue in the following theories. We will need to be selective in the items discussed, for the purposes of illustration or criticism. Bearing this in mind these sections cannot be taken as purporting to give a critical appreciation of the theories as a whole.

The previous section has been concerned with the existence of value as an order preserving homomorphism. The real utility of value theory arises through consideration of value structures and their contribution to computable problems and we discuss this in the next three sections.

To begin with, we need to realize that choice should be observed in the context of general, rather than dichotomous, problems. For this reason we consider some aspects of Milnor's theory. Although it takes a special form, the principle is valid for the remaining structures considered. Milnor's theory is essentially meant to be part of a theory of choice under uncertainty,[1] and we shall discuss it as such later on. At this stage this is not part of the point about value structures which we wish to make.

The essential point used in the next three sections is that: whereas χ was, in Section (2.2.1), merely any collection of elements, in these sections χ has a structure which is used in value determination. To put it briefly, we shall determine values of an element x in χ as a function of the values of the 'components' of x.

2.2.2.1 *Values and General Problems.* Milnor (51) takes as his format problems specified by a set of alternatives, a set of possible events,

---

[1] Baumol (6) defines situation as being uncertain if they are neither certain nor probabilistic, and, at this point, uncertainty is used in this sense. In contrast Chapter 3 uses uncertainty in the non-certain sense.

and a set of possible consequences. The problems can be varied by varying the set of alternatives, the set of exclusive and exhaustive events (hence changing the number) and varying the consequences. A typical format is

$$
\begin{array}{c}
\quad e_1\, e_2 \ldots e_j \ldots e_m \\
\begin{array}{c} a_1 \\ a_2 \\ \vdots \\ a_i \\ \vdots \\ a_k \end{array}
\left[\quad \theta(a_i,\, e_j). \quad\right]
\end{array}
$$

(a_i) are the alternative acts,
(e_j) are the possible events,

$\theta\,(a,\, e)$ is the value of taking act a when event e is realized.

Apart from varying the possible events and the set of alternatives the values[1] can be varied in a given manner, viz. by taking linear combinations. Milnor's axioms are concerned with the dependence of choice on variations in the form of the problem. Thus, to take an example, let us consider axioms 1 and 6.

Axiom 1: 'A transitive comparable order relation exists between the rows.'
Axiom 6: Row adjunction: 'The ordering between the old rows is not changed by the adjunction of a new row.'

If axiom 1 holds then axiom 6 must hold. The dependence of choice on the whole set of alternatives rather than on paired comparisons depends essentially on the violation of axioms 1 and 6.

Axiom 6 suggests that it might be possible, by adding alternatives, to change the solution, but to still retain this solution in the old problem alternatives set. Thus it might be possible that, given

$$
Q_1 \equiv \begin{bmatrix} a_1 \\ a_2 \\ a_3 \end{bmatrix} \text{ the choice is } a_1, \text{and given } Q_2 \equiv \begin{bmatrix} a_1 \\ a_2 \\ a_3 \\ a_4 \end{bmatrix} \text{ the choice is } a_2.
$$

The addition of a_4 has altered the solution from a_1 to a_2, but a_2 is still in Q_1. Axiom (6) removes this possibility.

Leaving aside the general validity of the axioms let us consider the

---

[1] From the point of view of the theory the values are simply numbers. However, in any practical use of this, we need to have a theory for getting the values $\theta(a, e)$.

value structures. Milnor proves that, if certain of the axioms are valid, choice is made in accordance with certain criteria. These are:

(i) Laplace: $\max_{a\varepsilon Q}[\sum_{e\varepsilon E}\theta(a, e)]$. (12)

(ii) Wald: $\max_{a\varepsilon Q}[\min_{e\varepsilon E}[\theta(a, e)]]$. (13)

(iii) Hurwicz (α): $\max_{a\varepsilon Q}[\alpha \max_{e\varepsilon E}[\theta(a, e)]$ (14)
$+ (1-\alpha) \min_{e\varepsilon E}[\theta(a, e)]]$.

(iv) Savage: $\max_{a\varepsilon Q}[\min_{e\varepsilon E}[\theta(a, e) - \max_{b\varepsilon Q}[\theta(b, e)]]]$. (15)

In order to see the difficulties which may arise when searching for value functions let us consider an application of Savage's criterion (15). Let Q_1, Q_2, Q_3 be defined as follows:

(16)

$Q_1 \equiv$

	e_1	e_2	e_3	e_4
A	2·5	2	0	1
B	1	1	2	1

$Q_2 \equiv$

	e_1	e_2	e_3	e_4
A	2·5	2	0	1
C	0	4	0	0

$Q_3 \equiv$

	e_1	e_2	e_3	e_4
B	1	1	2	1
C	0	4	0	0

The choice from Q_1 is B rather than A.
The choice from Q_2 is A rather than C.
The choice from Q_3 is C rather than B.
Thus axiom (1) is violated. Thus no value function for an alternative (independent of the problem) exists.

Another example, violating axiom (6), is given by the two problems Q_1, Q_2 specified below:

(17)

$Q_1 \equiv$

	e_1	e_2	e_3	e_4
a_1	$1\frac{1}{2}$	$1\frac{3}{8}$	$1\frac{1}{4}$	$\frac{3}{4}$
a_2	$\frac{1}{2}$	1	$1\frac{1}{8}$	$2\frac{1}{2}$
a_3	$\frac{1}{4}$	$\frac{1}{2}$	$\frac{3}{4}$	3
a_4	0	$\frac{1}{4}$	$\frac{1}{2}$	4

$Q_2 \equiv$

	e_1	e_2	e_3	e_4
a_1				
a_2		Q_1		
a_3				
a_4				
a_5	2	$1\frac{3}{8}$	$1\frac{1}{4}$	$\frac{1}{4}$

The choice from Q_1 is act (3). The choice according to Q_2 is act (2). Hence adding act (5) changes the choice, but retains it in the original act of acts (1)–(4). Hence again no value structure, independent of the problem, exists.

As a consequence of this, we may not be able to take values as an essential element in decision theory, unless we prohibit certain types of behaviour. Value theory centres around dichotomous type problems and around the transitivity property. The pairwise relation 'R' is indispensable to its validity. There need not, however, in general problem solving, be any form of pair wise comparison at all. Thus the most general relation is not

$$aRb \qquad\qquad (18)$$

but

$$aRb \text{ (rel } Q) \text{ for all } b \, \varepsilon \, Q. \qquad\qquad (19)$$

It makes no sense to have $aRbRc$ rel Q since we are asked to choose one alternative, a, from Q. Selecting a second alternative is then relative to the reduced problem $Q - a$.

Values perform a useful service to a theory of computable choice and having to bring in a relation 'R (relative to Q)' considerably complicates the matter.[1] Nonetheless, even without value structures, we do have general problem choice structures which serve much the same purpose, although to some extent complicating the issue.

General choice behaviours of types (i), (ii) and (iii) do have corresponding values for the acts.

The results only apply to the largest class of problems for which the axioms are valid. The real purpose of this theory lies in its application to choice under uncertainty. As a general theory it is meant to be applicable to situations of 'symmetry of information' about the events used in problem specification. Unfortunately this can usually only be obtained experimentally and in any actual business problem, one would need to have an operational definition of 'symmetry' to be verified by observing choice in certain situations.

One of Milnor's axioms (axiom 2) provides such an operational definition of 'symmetry'. (*Axiom 2*: The ordering of the rows is independent of the numbering of the columns.) This means that the value of an alternative does not depend on the relationship between

---

[1] The search for improvements on alternatives obtained to date has no meaning if value functions do not exist.

29

the events and consequences, but only on the spectrum of consequences for that alternative. This applies only when a value function exists. In the Savage minimax-regret situation, axiom 2 relates to each problem individually. This, in itself, does not rule out the probability approach, since the Laplace criterion is equivalent to one. It does, however, agree with Churchman's view that we only know what a person knows by observing his choice behaviour.

Although Milnor's theory can be tested within the context of specific problematic classes, the real intention is that, if axiom 2 is true, then we are in a situation of 'true' uncertainty, and one might expect that the criterion to be applied would then be independent of the problematic class; thus, in such situations, the decision-maker would use the same criterion whatever the physical nature of the decision situation, e.g. always a Laplacian.

2.2.2.2 *Value Structures in Dichotomous Choice Situations.* We shall now concern ourselves with situations in which choice is dichotomous, values exist, and in which values of elements can be shown to be computable functions of components values. These results are a mixture of applications in certain and uncertain situations, but this is of no differential importance within the context of the point of this section, viz. the computability of values.

In the case in which χ is an n dimensional Euclidean space over the real variable field, Ackoff introduces an 'additivity' axiom, as follows:

$$(x,y \, \varepsilon \, \chi) \rightarrow ([x+y \, \varepsilon \, \chi] \text{ and } [v(x+y) = v(x)+v(y)]). \qquad (20)$$

If the value structure is of this nature, then if $x_1, x_2 \ldots x_n$ form a basis for χ, and if $x = \Sigma \alpha_i x_i$, then

$$v(x) = \sum \alpha_i v(x_i). \qquad (21)$$

Once the $v(x_i)$'s have been obtained then any $v(x)$ can be 'decided'. To find these values choose $\lambda_1, \lambda_2, \ldots \lambda_n (= 1)$ so that

$$\lambda_1 x_1 \, I \, \lambda_2 x_2 \, I \ldots I \, \lambda_n x_n \qquad (22)$$

where 'I' means 'indifferent'.

By putting $v(x_n) = 1$ we get

$$\lambda_i v(x_i) = v(\lambda_i x_i) = v(\lambda_n x_n) = 1, \quad i = 1 \ldots n.$$

Thus:

$$v(x) = \sum \alpha_i \lambda_i^{-1}. \tag{23}$$

The axiom $(20)^1$ leading to (23) is an axiom about the values of composite elements. It is 'verifiable' only by observing that (23) predicts choice. It is not observable directly, since values are not directly observable.

Such an axiom serves a useful purpose in that, if it is valid, it is only necessary to store a limited number of such values, and the rest can be deduced. Hence they allow actions to be decided. A computer could now take over the job of choosing even if χ contained 10^{10} elements. This is well beyond its capacity to store the order propositions corresponding to each pair of elements in χ.

Let us now consider Koopmans' (42) approach to time series valuations. A time series is a vector of large dimension, whose components may also be vectors. $x = [x_1, x_2, \ldots x_n \ldots]$ is a time series where x_k may be a commodity consumption vector at time k.[2] It is impracticable to consider all paired comparisons of such vectors and it is only by the introduction of value structures that we can handle the problem of choice in dynamic situations.[3] Koopman provides such a value structure, dependent on certain axioms of choice. I will quote his first three axioms, which are the only ones relevant to the point at issue. The time series are assumed to be infinite.

Axiom 1. (Continuity). There is a utility function $U(_1x)$ which is defined for all $_1x = [x_1, x_2 \ldots x_n \ldots]$ such that, for all t, x_t is a point of a connected subset S of n dimensional commodity space. The function $U(_1x)$ has the continuity property that: if U is any of the values assumed by that function, and if U^1 and U^{11} are numbers such that $U^1 < U < U^{11}$, then there exists a positive number δ such that the utility $U(_1x^1)$, of every programme having a distance $d(_1x^1, _1x)$ for some programme $_1x$ with utility $U(_1x) = U$, satisfies $U^1 \leqslant U(_1x^1) \leqslant U^{11}$ where

---

[1] If a total order R exists, a necessary and sufficient axiom of choice is that if, x, $y, z, x+z, y+z \varepsilon \chi$, $xRy \rightleftarrows (x+z)R(y+z)$.

[2] The usual approach of constructing indifference surfaces is completely impracticable in situations with more than a small number of free variables. This is why the determination of value structures, via simple axioms, is useful, when feasible.

[3] This is connected with the Extensive approach to sequential decision-making (see Luce-Raiffa (45)).

$$d(_1x^1, _1x) = \sup_t \left[|_1x^1_t - _1x_t|\right] \tag{24}$$

Axiom 2. (Sensitivity). There exist first-period consumption vectors x_1, x^1_1 and a programme $_2x$ from the second period on such that:

$$U(x_1, _2x) > U(x^1_1, _2x) \tag{25}$$

Axiom 3. (Aggregation by Periods). For all x_1, x^1_1, $_2x$, $_2x^1$,

$$(U(x_1, _2x) \geqslant U(x^1_1, _2x)) \rightarrow (U(x_1, _2x^1) \geqslant U(x^1_1, _2x^1)) \tag{26}$$

$$(U(x_1, _2x) \geqslant U(x_1, _2x^1)) \rightarrow (U(x^1_1, _2x) \geqslant U(x^1_1, _2x^1))$$

Axioms (1), (2), (3), imply that there exist immediate utility functions $u_1(x_1)$, $u_2(x_2)$..., prospective utility functions $U_2(_2x)$, $U_3(_3x)$ ..., which have the strong continuity property of axiom (1), and a function $V(u, U)$, monotone increasing in u, U, such that

$$U(_1x) = V(u_1(x_1), U_2(_2x)) \tag{27}$$

Once V, u_1, U_2 have been determined, U can be computed, bearing in mind that, for a function of two variables with continuity and monotonicity properties, interpolation from a few points is easy. Alternatively we could express V, u_1, U_2 analytically. Of course U_2 would have to be determined in terms of u_2, U_3 and so on. Although these results are far from being able to tell us how to get the relevant functions, they are a step in the right direction. As we shall see later on (27) has a great deal of similarity to the dynamic programming approach to value determination, the essential difference being that V would be known, and we would bring in a decision variable as well. Under a given policy, for a deterministic system, we would have:

$$U_1(_1x) = V(x_1, x_2, U_2(_2x)). \tag{28}$$

So far we have been dealing with sets of elements χ which are vector spaces. In Ackoff's approach the composition property is via additivity of vectors, and values. In Koopman's approach the composition property lies in the relationships between a vector and a subvector and the values of each. von Neumann and Morgenstern (70) have considered sets, χ, whose elements are probabilistic combinations of other elements. There is nothing to prevent an element itself being a time series or a vector, but time series or vector characteristics do not come into the theory. This is not to say that the person concerned does not consider such characteristics or even 'decide' according to Ackoff's or Koopmans' valuation process. It is a theory of choice only based simply on probabilistic, and not vector, decomposition. Since von Neumann's theory is a central one in theories of choice under uncertainty the axioms are quoted. They are divided by Luce and Raiffa (45) into 'mixture' axioms, and 'order' axioms.

Mixture Axioms M.

Axiom 1.   For any $u,v \, \varepsilon \, M$ and any $p, 0 \leqslant p \leqslant 1$, $upv \, \varepsilon \, M$.

Axiom 2. $upv = v(1-p)u$.

Axiom 3. $up(vqw) = (u(p/(p+q-pq))v)(p+q-pq)w$
$\qquad\qquad (1-p)(1-q) \neq 1$

Order Axioms O.

Axioms 1 and 2 (combined).   '$>$' is a chain order (satisfies properties (i), (vi)*, (ix), (xii), of Section 2.1).

Axiom 3. $(u < v) \rightarrow (u < upv)$ $(p \neq 1)$

Axiom 4. $(u > v) \rightarrow (u > upv)$ $(p \neq 1)$

Axiom 5. $(u < w < v) \rightarrow (\exists p$ such that $upv < w)$

Axiom 6. $(u > w > v) \rightarrow (\exists p$ such that $upv > w)$

'$w \equiv upv$' has the interpretation that 'w is the element which is identical with element u or v with probabilities p and $(1-p)$ respectively'. Axiom (3) in the mixture axioms should be an order axiom since '$a = b$' is defined as '$a \not> b, b \not> a$' and these are essentially verifiable by choice. Thus axiom (3) says something about the equivalence of the prospects presented in different ways (see discussion further on).

If these axioms hold, then there exists a value function over M such that

$$V(apb) = pV(a)+(1-p)V(b) \qquad (29)$$

If U is any other value function satisfying (29), there exist a $\lambda \, (>0)$ and a μ such that

33

$$U = \lambda V + \mu \tag{30}$$

The axioms do not imply that V is the only value function, nor that all value functions are linear combinations of each other. They simply imply that a value function, satisfying (29), exists, and that (30) holds for other functions satisfying (29).

All of the axioms are relatively uncontroversial except for axiom $M3$ which requires that one compounds probabilities in the classical sense. More specifically, axiom $M3$ requires that all elements which can be reduced, using the probability calculus, to the same probability combination of specific set of elements, are equivalent from a choice point of view. Thus the elements $(a\xi b)p(a\eta b)$ and $(a(p\xi + (1-p)\eta)b$ are equivalent. This axiom is the one usually queried, for it is reasonable to expect that a person, who would be able to say that he was indifferent between the two, probably computes the probabilities for the second case before giving his judgement. If a person exhibits little consistent behaviour for this axiom, it may be because he has not evaluated, probability-wise, the form of his element in terms of the more basic elements. The problem is quite complex since it really needs a discussion of probabilities both as to the physical nature of their derivation and as to their 'sufficiency' as informational measures for choice, and this is taken up later on.

If one did compute probabilities of the basic elements, one would expect that every $u \varepsilon M$ would be reducible to a probability combination of a finite set of irreducible elements, and axiom $M3$ should not be questionable. However, axiom $M3$ says something about the manner in which the problem is constructed sequentially. First of all u is presented as an alternative outcome to (vqw). If u is realized we terminate the process. If u is not realized v is presented as an alternative outcome to w, and the process terminates when v or w is realized. The manner in which the right-hand side of axiom $M3$ is presented is entirely different to the manner in which the left-hand side is presented, and hence we might expect a difference in choice, unless the person actually compares them probabilistically, and bases his choice on the fact that both are equivalent, probabilistically, to alternative outcomes u, v, w with probabilities p, $(1-p)q$ and $(1-p)(1-q)$ respectively, and that, for him, probability is a jointly[1] sufficient informational measure for choice.

Let us return to our main theme concerning the relevance of this

[1] with value

theory for computable values. Computability lies in the value structure as exhibited by equation (29). This allows us to compute the values of complex probabilistic elements once the values of the basic elements have been obtained. Thus, if we were considering a very large number of alternative probability distributions over a set of monetary outcomes it would not be necessary to store the values of every possible probability distribution. Each value could be computed once the value structure of the monetary outcomes (in this case money) had been determined. Equation (29) thus results in an extensive widening of the size of problem which can be handled.

Misunderstandings of the purpose of determining value structures arises quite often in the literature on the subject. A typical misunderstanding is illustrated by a quotation from Baumol (5):

'More generally, where choice is concerned, two indices are genuine substitutes where they both lead [give the highest number] in all cases to identical decisions—as between buying and not buying . . . and, in this sense, any process such as the Neumann–Morgenstern construction, which limits the acceptable index to any narrower set [in this case those indices satisfying equation (29)] must establish arbitrary and unnecessary restrictions.'

This quotation emphasizes the apparent lack of differentiation between Sections (2.2.1) and (2.2.3) in this chapter. Von Neumann does not say his value function(s) are unique, only that some do exist which satisfy (29). Baumol gives three axioms of his own which effectively give the same valuation operations as von Neumann.[1] It is, however, entirely possible that von Neumann's key axiom, $M3$, may be violated, and hence (29) may not be valid. Baumol's results only give an order preserving homomorphism, with no usable structural properties, and hence of no value in making large classes of problem decidable, which may otherwise be beyond the bounds of computation.

A further criticism, levied by Davidson (22), is that axiom $M1$ requires M to contain an infinite number of elements, whereas, quite obviously such value structures do exist for a finite number of elements. Thus if $M \equiv (E, F, E(\cdot 5)F)$ and if $E > E(\cdot 5)F > F$ if we put $v(E) = 1$, $v(F) = 0$, $v(E(\cdot 5)F) = \cdot 5$ this satisfies equation (29).

---

[1] These are equivalent to finding the probability combination of two extreme elements, which is indifferent to the given element, and equating the value of the given element to this probability.

Davidson correspondingly constructs a theory in which M can be finite. Even if M is not infinite, if we emphasize that value theory has real usefulness only when M is at least large enough to prohibit the straight storage of paired comparisons, it might just as well be taken as infinite. We would like to be able to cater for any reasonable possibility arising in complex situations. Thus, in the above problem it would seem reasonable to want a value structure which would include EpF for all $0 \leqslant p \leqslant 1$, even though value structures can be determined over a finite set allowing the relaxation of certain axioms.

Let us consider, for example, a general dynamic problem of valuing the state of a system, subject to probabilistic transitions and operating under a given policy over some time period. Suppose that only the final state of the system matters (e.g. total cash at the end). Using von Neumann's results, Smith (65) derives the Kolmorogov–Chipman value equation.

$$v(x, s) = \sum_y p(x, s; y, t)\, v(y, t) \tag{31}$$

where $v(x, s)$ is the value of being in state x at time s, $p(x, s; y, t)$ is the probability that, if we are in state x at time s, we shall be in state y at time t.

The number of possible states, particularly if x is a vector, can be colossal. Also the number of probability distributions could be very large if we consider a wide variety of different policies. We obviously need a theory which covers as large an M as possible and it certainly needs to cover many probability combinations over M, where M is the set of all pairs (x, s) in some region.

It is worth noting that by removing the Archimedian type axioms (O5 and O6), Hausner (36) shows that von Neumann's results can be generalized to lexicographic values, extending Chipman's results to probabilistic situations.

Computations with lexicographic values follow similar lines to von Neumann's, component by component. Thus if $[p_1, p_2 \ldots p_m]$ $[p_1^1, p_2^1 \ldots p_m^1]$ are two lexicographic values, corresponding to elements $u, u^1 \, \varepsilon \, M$ respectively, then the element uqu^1 has lexicographic value

$$[qp_1 + (1-q)\, p_1^1,\ qp_2 + (1-q)\, p_2^1, \ldots qp_m + (1-q)p_m^1].$$

Savage (60) produces a value structure identical to that of von

Neumann, but in this case the probabilities of outcomes are not specified. The problematic set up is similar to that of Milnor in which events are specified but not their probabilities. His set χ is a set of acts, F.[1] There exists a set of states S, forming a Boolean algebra of classes. There exists a set of consequences, C. Any act, $\underline{f} \varepsilon F$, is then determined by allocating a consequence, $f \varepsilon C$, to each state $s \varepsilon S$. Each consequence, f, is identified with an act $\underline{f}$ whose allocation is f to each state s. The physical interpretation is that if act $\underline{f}$ is taken, and if state s is realized, then the outcome is $f(s)$.

As with von Neumann a consequence is, by identification with an act, to be considered as an act itself. Savage's consequences and states are, by definition, not reducible. Savage's theory is not, therefore, applicable to sequential decision-making, although, with due modification, it is.[2]

Savage's theory holds a central position in decision-making under uncertainty.[3] For this reason we quote the axioms in full both for present use (in the context of value theory) and for future use (in the context of uncertainty theory).

Axiom 1.

The relation $\leqslant$ is a simple[4] ordering among acts $\underline{f} \varepsilon F$ (satisfies properties (i) and (vi) of Section (2.1)).

Axiom 2.

If $\underline{f}, \underline{g}, \underline{f}', \underline{g}'$ satisfy:
1. in $\sim B, \underline{f}$ agrees with $\underline{g}$ and $\underline{f}'$ agrees with $\underline{g}'$
2. in $B, \underline{f}$ agrees with $\underline{f}', \underline{g}$ agrees with $\underline{g}'$
3. $\underline{f} \leqslant \underline{g}$,
   then $\underline{f}' \leqslant \underline{g}'$.

Axiom 3.

If $\underline{f} \equiv f$ (constant act), $\underline{g} \equiv g$ and B is not null (virtually impossible) then $\underline{f} \leqslant \underline{g}$ given B if and only if $\underline{f} \leqslant \underline{g}$. ('$\underline{f} \leqslant \underline{g}$ given B' is to be

---

[1] We will use Savage's own notation.

[2] In order to do this we need to be able to cater for the conjunction of states of nature and also to make $F \equiv C$, so that any act in F is also a consequence in C.

[3] Here we use 'uncertainty' in the sense of 'non-certainty'.

[4] A simple order is virtually the same as a total order. In Savage's theory, axiom 5 is the same as condition (x) in Chipman's relations.

37

interpreted as 'if f and g are modified so that they agree on $\sim B$. then f would be preferred to g if B were known.' This is part of Savage's 'sure thing' principle.)

Axiom 4.

If $f, f', g, g', A, B, \underline{f}_A, \underline{f}_B, \underline{g}_A, \underline{g}_B$ satisfy

1. $f' < f, g' < g$
2a. $\underline{f}_A(s) = f, \underline{g}_A(s) = g, s \varepsilon A$
    $\underline{f}_A(s) = f', \underline{g}_A(s) = g', s \varepsilon \sim A$
2b. $\underline{f}_B(s) = f, \underline{g}_B(s) = g, s \varepsilon B$
    $\underline{f}_B(s) = f', \underline{g}_B(s) = g', s \varepsilon \sim B$
3. $\underline{f}_A \leqslant \underline{f}_B$

then $\underline{g}_A \leqslant \underline{g}_B$.

Axiom 5.

There exists at least one pair of consequences f, f' such that $f < f'$.

Axiom 6.

If $B < \cdot C$, then there exists a partition of S, the union of each element of which with B is less probable than C. ('$B \leqslant \cdot C$' is interpreted as 'B is not more probable than C', and is equivalent, operationally, to 'if $f' < f$, and $\underline{f}_B, \underline{f}_C$ are acts such that:
    1. $\underline{f}_B(s) = f$ for $s \varepsilon B, \underline{f}_B(s) = f'$ for $s \varepsilon \sim B$

    2. $\underline{f}_C(s) = f$ for $s \varepsilon C, \underline{f}_C(s) = f'$ for $s \varepsilon \sim C$
then $\underline{f}_C \leqslant \underline{f}_B$.' If $B \leqslant \cdot C$ but $C \not\leqslant \cdot B$, then $B < \cdot C$.)

Axiom 6^1.

If $g < h$ and f is any consequence then there exists a partition of S such that if g or h is so modified on any one element of the partition as to take the value of f at every s there, other values being undisturbed, then the modified g remains less than h or h remains less than the modified g, as the case may require.

Axiom 7.

Only a finite number of consequences, f, exist.

Axiom 8.

If $f \leqslant (\geqslant) g(s)$ given B for every $s \, \varepsilon \, B$ then $f \leqslant (\geqslant) g$ given B.

Axioms (1)–(7) imply that there exists an order preserving homomorphism, v, on the acts, F, with respect to the order relation $\geqslant$ on F, (and this includes the consequences, C, interpreted as equivalent constant acts). There also exists a homomorphism,[1] p, on the Boolean algebra of sets of S such that v, p have the structural properties:

$$v(f) = \sum_{s \varepsilon S} p(s) v(f(s)) \tag{32}$$

$$p(A \cup B) = p(A) + p(B) \text{ if } A \cap B = \phi \tag{33}$$

$$\sum_{s \varepsilon S} p(s) = 1. \tag{34}$$

Axiom (8) simply allows extension to the case when C contains an infinite number of elements.

As far as the axioms are concerned, any critical appreciation must be left outside the context of this report. We can make one or two comments, however. It is reasonably obvious that the theory was probably developed so that a required result would be achieved. This result is the validity of the 'expected value' criterion as a value function. Axiom (4) is particularly strong in this direction.

Consider the following problem, specifying the consequences of acts f_A, g_A, f_B, g_B if certain events A, $\widetilde{A}$, B, $\widetilde{B}$, occur.

	A		$\widetilde{A}$	
f_A	f	f	f'	f'
g_A	g	g	g'	g'
f_B	f'	f	f	f'
g_B	g'	g	g	g'
	$\widetilde{B}$	B		$\widetilde{B}$

If the expected value function were valid we would get:

---

[1] p is actually a 'probability' function preserving the order of an induced qualitative probability relation $\geqslant$. This is discussed later on in the chapter on probability where it rightly belongs.

$$v(\underline{f}_A) - v(\underline{f}_B) = (v(f') - v(f))(p(B \cap \tilde{A}) - p(A \cap \tilde{B}))$$

$$v(\underline{g}_A) - (v(\underline{g}_B) = (v(g') - (v(g))(p(B \cap \tilde{A}) - p(A \cap \tilde{B}))$$

Thus, under the conditions of axiom 4, $\underline{f}_A \geqslant \underline{f}_B$ iff $\underline{g}_A \geqslant \underline{g}_B$, the ordering depending only on the sign of some element $p(B \cap \tilde{A}) - p(A \cap \tilde{B})$. Axiom (4) is thus a very strong axiom which may be refuted.

The axiom which is supposed to be beyond question is the so-called 'sure thing' axiom. Diagrammatically it can be represented as follows, where common signs denote identical consequences for a given state.

	B	$\tilde{B}$
$\underline{f}$	$0_1 \; 0_2 \ldots 0_m$	$x_1 \; x_2 \ldots x_n$
$\underline{g}$	$\Delta_1 \; \Delta_2 \ldots \Delta_m$	$x_1 \; x_2 \ldots x_n$
$\underline{f}'$	$0_1 \; 0_2 \ldots 0_m$	$v_1 \; v_2 \ldots v_n$
$\underline{g}'$	$\Delta_1 \; \Delta_2 \ldots \Delta_m$	$v_1 \; v_2 \ldots v_n$

The requirement that 'if $\underline{f} \leqslant \underline{g}$, then $\underline{f}' \leqslant \underline{g}'$' seems plausible enough. However, it may only be plausible to someone steeped in probabilities. As we have seen when considering Milnor's theory, a person may pay attention solely to the worst possible outcome of an act. Consider the following example.

	B	$\tilde{B}$
$\underline{f}$	·5	0
$\underline{g}$	1	0
$\underline{f}'$	·5	2
$\underline{g}'$	1	2

A maximiner would be indifferent between $\underline{f}$ and $\underline{g}$, but not between $\underline{f}'$ and $\underline{g}'$.[1] Of course one could say that if two minimal

---

[1] A consequence of axiom (2) is that $\underline{f} = \underline{g}$ implies $\underline{f}' = \underline{g}'$.

components are equal then we use the second component. Then we have lexicographic values and not real values, and Savage's theory falls down. One must admit that the sure thing principle seems beyond reasonable dispute, and it is likely that the maximin principle is invalid.

In order to be able to compute the values of complex elements we need to be able to determine $v(f), f \varepsilon C$, and $p(s), s \varepsilon S$. One way of doing this is to calculate $p(s), s \varepsilon S$, first of all, and this is discussed in the relevant section on uncertainty. $v(f)$ is then calculated by finding an f_e, f_u, such that $f_e \leqslant f \leqslant f_u$, putting $v(f_e) = 0, v(f_u) = 1$ and then finding an $s \varepsilon S$, where $f_u s f_e = f$.[1] Then $v(f) = p(s)$.

An alternative approach lies in the possibility of being able to determine $[p(s)], [v(f)]$ analytically, in the case when S, F, C are finite. Each pair of equivalent acts, say $f \equiv (f(1), f(2) \ldots f(m))$ and $g \equiv (g(1), g(2) \ldots g(m))$, give rise, through (32), to an equation of the form

$$\sum_s p(s) (v(f(s)) - v(g(s))) = 0. \tag{36}$$

If there are m states, and N different consequences with $f_1 > f_2 > \ldots > f_N$, we have N^m possible acts. We have $(m + N)$ variables to determine out of which two can be given values 1 and 0 respectively. We also have

$$\sum_1^m p_s = 1.$$

Hence, if we can determine $(m + N - 3)$ equations of the form (36), we can solve for the required variables. As an example, consider the following situation, with $c_1 \neq c_2$:

$C \equiv [c_1, c_2], S \equiv [s_1, s_2], F \equiv [[c_1, c_1], [c_1, c_2], [c_2, c_1], [c_2, c_2]]$. Suppose $[c_1, c_2] = [c_2, c_1]$. We then have:

$$p_1(v(c_1) + p_2 v(c_2) = p_1 v(c_2) + p_2 v(c_1). \tag{37}$$

Putting $v(c_2) = 0, v(c_1) = 1$ we get $p_1 = p_2 = \frac{1}{2}$.

As another example, let $C = [c_1, c_2, c_3], S = [s_1, s_2]$.
$F = [[c_1, c_1], [c_1, c_2], [c_1, c_3], [c_2, c_1], [c_2, c_2], [c_2, c_3], [c_3, c_1], [c_3, c_2], [c_3, c_3]]$.
Let $v(c_1) > v(c_3)$.

---

[1] This is interpreted as the act whose consequence is f_u if s occurs, and whose consequence is f_e if s does not occur.

Suppose $[c_1, c_2] = [c_2, c_3], [c_1, c_3] = [c_3, c_2].$
We then have:

$$p_1 \, v(c_1) + p_2 \, v(c_2) = p_1 \, v(c_2) + p_2 \, v(c_3). \tag{38}$$

$$p_1 \, v(c_1) + p_2 \, v(c_3) = p_1 \, v(c_3) + p_2 \, v(c_2). \tag{39}$$

Putting $v(c_1) = 1$, $v(c_3) = 0$, we get $p_1 = \frac{2}{3}$, $p_2 = \frac{1}{3}$, $v(c_2) = 2$.

If, in the above examples, we now had 1,000 possible consequences, then we would have 10^6 possible acts, all of whose values are computable if we can isolate 999 equivalent pairs.[1]

So far we have dealt with value structures when choice is consistent and involves no random behaviour. Let us now consider a value structure under conditions of probabilistic choice, remembering that it is the value aspect with which we here are concerned and not the probabilistic aspect.

Luce (45) extends Debreu's work (24) to the case when an act may be a composite of two consequences, the actual consequence depending on whether a certain event α does or does not occur. Thus '$a = b\alpha c$' is to be interpreted as 'a is the act whose consequence is b if α occurs, and c if $\tilde{\alpha}$ occurs'. The axioms Luce uses are complex, and are, as he points out, inconsistent. However, since we are solely concerned with value structures, we quote the results for probabilistic choice in the hope that the inconsistency may later be removed.

The basic premises are as follows. Given acts a and b, a is chosen with probability, $P(a, b)$. Given two exclusive events (not necessarily exhaustive) α, β, there exists a probability function, $Q(\alpha, \beta)$, which is the probability that the person would say that α would occur when he has to choose between α and β. He defines $a \geqslant b$ iff $P(a, b) \geqslant \frac{1}{2}$, and $\alpha \geqslant \beta$ iff $Q(\alpha, \beta) \geqslant \frac{1}{2}$. He postulates that order preserving homomorphisms, $u(a)$, $\phi(\alpha)$, exist.[2] He then postulates the validity of eleven axioms of choice. We quote one of these for future use, viz:

$$u(a\alpha b) = \phi(\alpha)u(a) + \phi(\tilde{\alpha})u(b) \quad (\tilde{\alpha} = \text{not } \alpha). \tag{40}$$

He establishes the existence of a positive ε such that

---

[1] We could deal with inequalities in (36) instead of equalities, but this would only limit $p(s)$, and $v(f)$ to given regions. This might, however, give an adequate approximation.

[2] u is a value, ϕ is a subjective probability.

$$Q(\alpha, \beta) = \begin{cases} \frac{1}{2} + \frac{1}{2}(\phi(\alpha) - \phi(\beta))^\varepsilon & \text{if } \alpha \geqslant \cdot\beta \\ \frac{1}{2} & \text{if } \alpha = \cdot\beta \\ \frac{1}{2} - \frac{1}{2}(\phi(\beta) - \phi(\alpha))^\varepsilon & \text{if } \alpha \leqslant \cdot\beta \end{cases} \tag{41}$$

$$P(a, b) = \begin{cases} \frac{1}{2} + \frac{1}{2}\lambda(u(a) - u(b))^\varepsilon & \text{if } a \geqslant b \\ \frac{1}{2} & \text{if } a = b \\ \frac{1}{2} - \frac{1}{2}\lambda(u(b) - u(a))^\varepsilon & \text{if } a \leqslant b \end{cases} \tag{42}$$

where

$$\lambda = P(a^*, b^*) - P(b^*, a^*) \neq 0$$

for some a^*, b^*.

Once ε has been determined the subjective probability of α can be determined from (41) by putting $\tilde{\alpha} = \beta$, using $\phi(\alpha) + \phi(\tilde{\alpha}) = 1$ (this follows from (40) by putting $a \equiv b$, and assuming $0 \not\equiv u(a)$ for all a), and observing $\phi(\alpha, \tilde{\alpha})$. $u(a\alpha b)$ can then be computed from (40) once $u(a)$, $u(b)$ have been determined for basic components a, b. Since we can put $u(b) \equiv 0$ for some $b \equiv b'$ say, $u(a)$ can be computed from (42) by observing $P(a, b')$ for the specific b' for which $u(b')$ is taken equal to 0. Then $P(a, b)$ can be computed from (42) for all a, b. Hence from the consideration of a few events and a few acts we can compute the remaining elements.

The calculation of ε might proceed as follows. Let $u(a') = 1$, $u(b') = 0$. This is possible since $\gamma u + \mu$ is a permissible value scheme if u is permissible, if $\lambda > 0$. Then $u(a'\alpha b') = \phi(\alpha)$ (from (40)). Hence, from (42), if $a'\alpha b' \geqslant b'$, $P(a'\alpha b', b') = \frac{1}{2} + \frac{1}{2}\lambda(\phi(\alpha))^\varepsilon$ (if $a'\alpha b' \leqslant b'$, similar results can be derived). Now choose α so that $Q(\alpha, \tilde{\alpha}) = \frac{1}{2}$. Then $P(a'\alpha b', b') = \frac{1}{2} + \frac{1}{2}\lambda(\phi(\alpha))^\varepsilon = \frac{1}{2}(1 + \lambda 2^{-\varepsilon})$ and if $P(a'\alpha b', b')$ is observed we can compute ε (λ is already known once a^*, b^* have been selected).

This section has been concerned with the possibility of being able to deduce values, and hence choice, in specific circumstances, from consideration of only a relatively few choice situations. It is seen that the axiomatic approach is indispensible to a computable theory of value in situations where a large number of alternatives and consequences are present and prevent direct enumeration of each element and coverage of all paired comparisons. Essentially, as in mathematical analyses, the value structures considered introduce an element of decidability into choice.

There remains, of course, the problem of axiom verification. This may be done on an 'ought' basis by accepting the axioms as a postulates of universally desired behaviour. Alternatively it is simply a question of verifying that the axioms are actually adhered to in a finite selection of relevant choice situations. If they do not describe such choice behaviour correctly we face the problem of ascribing such deviations to errors of several kinds and whether or not to accept the axioms as approximately correct. A discussion of this, while within the compass of this report, must be left to some future date, since it obviously involves severe difficulties arising from the values of theories in application.

We have discussed value structures in general. Let us consider the special case of sequential processes to which von Neumann's work has great relevance. It is in sequential processes that value structures have their greatest value.

2.2.2.3 *Value Structures and Sequential Problems.* Savage's theory deals with acts, states of nature and 'consequences'. The determination of appropriate consequences with which to define the outcome of an act is a central problem in decision theory, although it is by no means necessary that a person should ever consider consequences when selecting an action. The aspect of consequence representation we shall deal with concerns the decomposition of the consequences of an act. Decomposing[1] a consequence introduces a sequential character to the problem.

Savage queries whether consequences can ever be reduced to a composite mixture of indecomposable consequences. All consequences have, themselves, consequences. Money may be decomposable if it is used for some purpose. If we purchase a motor car, this is decomposable, for we shall use it to carry out certain activities. These activities may result in other consequences. Thus if we crash, we spend a certain time in hospital, and this has consequences.

A computable theory of decision requires that the consequence tree be terminated after a finite number of steps, and this may require an artificial termination. Thus, in Operational Research, 'quality' has sales consequences, but we have, as yet, no satisfactory way of determining them.

When dealing with sequential processes there are two possible

---

[1] Decomposition will, for our purposes, relate to the possibility of breaking down a consequence and its value, into further consequences and their values. It has, therefore, a restricted meaning in this book.

points of view [1] depending on whether we are concerned with the choice of a decision rule for the sequential process or with choice of an individual act at any one stage. The two possible formats are given below (see Raiffa and Schlaiffer (57)).

NORMAL FORM

A decision rule is considered as an alternative action, and the problem is one of choosing a best decision rule over a sequence of time periods.

EXTENSIVE FORM

At each action epoch, the action to be taken at that epoch is considered individually, and the optimal such action for the remaining periods is required. [2]

The difficulty with the Normal form lies in a relative inability to describe, in advance, all the possible decision rules. The difficulty with the Extensive form lies in the problem of determining the consequences of an action at any given stage.

The decision formats would be as follows:

	Alternatives	States of Nature	Consequences
Normal	Decision rules	Time series of uncontrollable events	Time series of outcomes
Extensive	Action at a given stage	Event in next stage	?

The question mark in the Extensive form concerns the form the consequences take in representing the residual of the sequential problem, once an action has been selected.

Let us consider the Normal form. A person chooses between acts (policies) on the basis of propositions about time series generated by the acts. Thus if an investment policy is specified and a probabilistic environment is assumed valid, for each policy we can make probability statements about time series of cash flows. The choice is then based on such probability statements. However, the number of

---

[1] The following discussion relates to the importance of value structures in sequential decision-making from the Extensive point of view.

[2] Thus, in the acts—states of nature—consequences format, the consequences of the individual decision at a given time are the total effect of this single decision taken in the light of future decisions.

policies which are possible could be colossal. If the policy is assumed to be stationary with respect to time, if there are D possible acts for every state, and if there are S states, the number of complete policies is D^S. $D = 10$, $S = 10$ gives 10,000,000,000 possible policies.[1] Such a problem is small when compared with many realistic problems. If the policy depends on the time of action, the computational problem is even worse.

The theory of value can be used here by utilizing the Extensive form together with existing value structures as in Section (2.2.2.2). Let a be an initial act. Let s be the state variable which, together with the act a and the policy[2] π, determines the consequence, θ, of taking act a when s is the true state of nature at the first stage. θ may be a vector such as $\theta(a, s) \equiv (c(a, s), p_{11}(a, s) \ldots p_{nm}(a, s))$ where $c(a, s)$ is immediate cost and $p_{ij}(a, s)$ is probability that the cost in the ith period will be c_j. These are all determined by a, s. We assume that there are n stages and D possible acts at each stage.

Let A be the set of acts (a). Let Θ be the set of consequences (θ). Each act a is then described by a vector,

$$a = [\theta(a, 1), \theta(a, 2) \ldots \theta(a, m)] \tag{43}$$

Suppose now we have a value function (real), v_m, relative to S over the space Θ^m ($A \subseteq \Theta^m$ quite obviously). Suppose v_1 is a value function over Θ. v_m will depend on the state structure for the first stage. v_1 will depend only on the form of the consequence space Θ.

Now it has already been stated (Section (2.2)) that value is simply an order preserving homomorphism over a space χ of alternatives. Hence v_1, in particular, may have little informational content for choosing an act a. Thus let us consider the following example, in which the matrix entries are values for the (sequential) consequences of the initial act.

$$\tag{44}$$

		s_1	s_2	s_3
$Q \equiv$	a	0	0	10^6
	b	1	1	1

---

[1] A complete policy specifies, for each condition, which act to take; a partial policy specifies, for each condition, a set of possible acts to take.

[2] We refer to the application of π to the remaining stages of the sequential process. This we assume fixed, leaving a, only, open to choice.

If we now allow any order preserving homomorphism of the consequences of the initial acts, then the following problem ought to give the same choice.

$$Q \equiv \qquad (45)$$

		s_1	s_2	s_3
	a	0	0	1·1
	b	1	1	1

[1]Order preserving homomorphisms for consequences are not necessarily sufficient variables for sequential processes, if such a homomorphism is to be used directly as an argument in the criterion for evaluating the initial decision.

If such a homomorphism were to be sufficient, it ought to be derived within the context of problem Q and not outside it. The valuation procedures for Θ and Θ^m ought to be connected, relative to S.

Von Neumann overcomes this difficulty since, if χ is a set of elements to be valued, and if $x = upv \ \varepsilon \ \chi$, then $u,v \ \varepsilon \ \chi$. In other words the consequence of an act are themselves considered as acts within the same value framework.

If von Neumann's valuation procedure were valid we would then obtain, for a process whose performance depends on total cost

$$v_m(a) = \sum_s p(s) \, v_1(\theta(a, s))$$

$$v_1(\theta(a, s)) = \sum_{j_1 j_2} \prod_i p_{ij_i}(a, s) \, v\left(c(a, s) + \sum_i c_{j_i}\right) \qquad (46)[2]$$

where $v(z)$ is the value of a final total cost z

$p(s)$ is the probability that state s will arise at the end of the first stage.

This still leaves the problem of getting the $p_{ij}(a, s)$'s. A complete model could involve the use of Dynamic Programming. However, the point we are trying to make here is simply that value structures can be used to resolve sequential problems using the Extensive form.

[1] One way out of this is to find comprehensible consequences with the same values and to represent the outcomes of these. Thus if $v_1(\theta) = v_1(\pounds10,000)$, then we could replace θ by $\pounds10,000$. Doing this for all outcomes reduces the problem to a simple form.

[2] For purposes of illustration we assume that $c_{j_1}, c_{j_2}, \ldots c_{j_n}$ form an independent set, but of course they need not do so.

In the case of non-probabilistic sequential processes Koopman's equation (27) may be of use, once the consequence function, $_2U$, has been determined. This depends on a consequence function $_3U$ and so on. For cases when stationary consequence functions exist[1] (27) reduces to:

$$U(x_1, x) = V(u(x_1), U(x)) \qquad (47)$$

Consequences are treated within the same framework as acts. x is considered both as an act and a consequence.

The dangers of using different valuation procedures for acts and consequences have been illustrated by equations (44) and (45). The dangers we run are further illustrated by Ackoff (2) in which he mixes two valuation procedures. For each act, i, there is a probability vector $p(i)$, where $p(i,j)$ is the probability that act i will result in outcome θ_j, θ_j is a vector. A valuation procedure, u, for the (θ_j) alone, is used, based on his additivity assumptions. The final choice is made on the basis of expected values using von Neumann's value structure, in the following form:[2]

$$v(i) = \sum_j p(i, j) u(j) \qquad (48)^2$$

Even if the $u(j)$'s were uniquely determined by the procedure suggested, we would have little reason to accept (48) as valid, because if u is a value function, then so is u^2. The use of u^2 will certainly change the ordering of the $v(i)$'s in certain cases.

In this case the specified procedure may not even give unique values of u. To illustrate the misuse of valuation processes of this kind, consider the following example, additivity being assumed.

$$\theta(1) = [0, 0], \theta(2) = [1, 1], \theta(3) = [1, 0], \theta(4) = [0, 1]. \qquad (49)$$

Let

$$\begin{array}{ll} u(\theta(1)) = 0, & u(\theta)(2)) = u(\theta(3)) + u(\theta(4)), \\ u(\theta(3)) = u(1), & u(\theta(4)) = u(2) \end{array} \qquad (50)$$

---

[1] A stationary function is one independent of the time at which it is evaluated.

[2] Where $u(j)$ is shorthand for $u(\theta_j)$.

[3] Thus if $\theta \equiv [x_1, x_2, \ldots x_n]$, then using Ackoff's additivity property (if it holds) we get $u(\theta) = \Sigma \lambda_i x_i$; using von Neumann's approach, if it holds, we get another value function $u^*(\theta)$. Even if we manipulate the degrees of freedom inherent in the scales and origins of both, there is no reason why $u(\theta)$ should be identical with $u^*(\theta)$.

Suppose, for a given policy, $\pi(e)$, the probabilities for $(\theta(i))$ are, respectively $p(e)$, $q(e)$, $r(e)$, $t(e)$.

Then, if $v(e)$ is value of $\pi(e)$

$$v(e) = (q(e)+r(e))\,u(1)+(q(e)+t(e))u(2) \tag{51}$$

then $v(1) \underset{\leqslant}{\overset{\geqslant}{}} v(2)$ iff

$$(q(1)+r(1)-q(2)-r(2))\,u(1)+(q(1)+t(1)-q(2)-t(2))\,u(2) \underset{\leqslant}{\overset{\geqslant}{}} 0 \tag{52}$$

Let $q(1) = q(2)$; $r(1)-r(2) = {\cdot}2$, $s(1)-s(2) = -{\cdot}1$.

Then $v(1) \underset{\leqslant}{\overset{\geqslant}{}} v(2)$ iff

$$2u(1) \underset{\leqslant}{\overset{\geqslant}{}} u(2). \tag{53}$$

Now suppose $\theta(3) < \theta(2)$, and hence $u(1) < u(2)$, then $u(1) = 2$, $u(2) = 3$ or $u(1) = 1$, $u(2) = 3$ would satisfy these requirements. However, the former valuation would give $2u(1) > u(2)$ (leading to choice of $\pi(1)$) and the latter valuation would give $2u(1) < u(2)$ (leading to choice of $\pi(2)$).

Admittedly the example is simple, but the same applies to situations where the consequences are multidimensional vectors over E_n. To make sure that we are not violating von Neumann's approach[1] the values $(u(\theta u))$ ought to be obtained in the same way as the $v(j)$'s. Thus, to get $u(1)$ we would let $u(\theta(1)) = 0$, $u(\theta(2)) = 1$ and then find p so that $\theta(3) \equiv \theta(1)p\theta(2)$. Then $u(1) = u(\theta(3)) = (1-p)$.

Although the valuation of consequences and acts using the same procedure is valid in von Neumann's theory, it must not be assumed that this will be the case for all sequential situation. If we are not careful we can get contradictory results for some theories of choice.

Consider, for example, Hurwicz's α criterion as given by equation (14). Although this valuation method may result in an order preserving homomorphism for the acts in a non-sequential problem, we cannot use this valuation process in a two stage problem for valuing the second stage consequences. Consider the following problem. The first stage is represented thus:

---

[1] Assuming, of course, that von Neumann's theory is valid over $A \cup H$, where A is the set of alternatives, and H is the set of consequences.

state j

$$Q \equiv act: \quad \begin{array}{c|c|c} & Q(1,\,1) & Q(1,\,2) \\ \hline 2 & Q(2,\,1) & Q(2,\,2) \end{array}$$

The consequence of taking act i when the state (first stage) is j, is a problem $Q(i, j)$.[1]

Let $Q(i, j)$ be given as follows:

(55)

state k   *state k*
1   2   1   2

$$Q(1, 1) \equiv \begin{array}{c|c|c} 1 & 1 & 1 \\ \hline 2 & 2 & 0 \end{array} \qquad Q(1, 2) \equiv \begin{array}{c|c|c} 1 & 0 & 0 \\ \hline 2 & 0 & 0 \end{array}$$

$$Q(2, 1) \equiv \tfrac{3}{4} \qquad\qquad\qquad Q(2, 2) \equiv \tfrac{3}{4}$$

Let us consider the Extensive form using Hurwicz α value function with $\alpha = \tfrac{1}{2}$.

(i) If we value the $Q(i, j)$ using this function we get

(56)

state j
1   2

$$Q \equiv \begin{array}{c|c|c} 1 & 1 & 0 \\ \hline 2 & \tfrac{3}{4} & \tfrac{3}{4} \end{array}$$

Using the Hurwicz criterion again we obtain the choice, in Q, of act 2.

(ii) Suppose we had used the Hurwicz α criterion, not to value $Q(i, j)$ but to determine what alternative would be selected for each $Q(i, j)$. The selections would have been either 1 or 2 in each case. We assume that he chooses alternative 1 in both cases.

---

[1] Thus we may have to choose between producing a product at one level or another. When the demand is known (stage j), at the end of the first year, we will be faced with another problem of whether to continue or not. The ensuing outcome depends on the next demand and the decision to continue or to discontinue.

Problem Q then reduces to the following form, in terms of the combined states, (j, k).

$$
\begin{array}{c}
state\ (j, k) \\
(1, 1)\ (1, 2)\ (2, 1)\ (2, 2)
\end{array}
\qquad (57)
$$

$$
Q \equiv
\begin{array}{c|c|c|c|c|}
 & (1,1) & (1,2) & (2,1) & (2,2) \\
\hline
1 & 1 & 1 & 0 & 0 \\
\hline
2 & \frac{3}{4} & \frac{3}{4} & \frac{3}{4} & \frac{3}{4} \\
\hline
\end{array}
$$

If he now uses the Hurwicz $\alpha = \frac{1}{2}$ criterion, we see that he chooses act 2 in Q. If he had chosen act 2 at the second stage, instead of act 1, Q then becomes

$$
\begin{array}{c}
state\ (j, k) \\
(1, 1)\ (1, 2)\ (2, 1)\ (2, 2)
\end{array}
\qquad (58)
$$

$$
Q \equiv
\begin{array}{c|c|c|c|c|}
 & (1,1) & (1,2) & (2,1) & (2,2) \\
\hline
1 & 2 & 0 & 0 & 0 \\
\hline
2 & \frac{3}{4} & \frac{3}{4} & \frac{3}{4} & \frac{3}{4} \\
\hline
\end{array}
$$

The Hurwicz $\alpha = \frac{1}{2}$ choice for Q is then act 1.

(iii) Let us now consider the Normal form of Q.

A decision rule specifies the first action, and which action to take when the first state j is known. Since there are no alternative actions for $Q(2, 1)$ and $Q(2, 2)$, there are five decision rules and the outcomes, in terms of the combined state (j, k), are given as follows:

	Decision Rule		state (j, k)			
First Action	Action if $j = 1$	Action if $j = 2$	$(1, 1)$	$(1, 2)$	$(2, 1)$	$(2, 2)$
1	1	1	1	1	0	0
1	1	2	1	1	0	0
$Q.$ 1	2	1	2	0	0	0
1	2	2	2	0	0	0
2	—	—	$\frac{3}{4}$	$\frac{3}{4}$	$\frac{3}{4}$	$\frac{3}{4}$

The Hurwicz $\alpha = \frac{1}{2}$ solution is then either to choose decision rule $(1, 2, 1)$ or $(1, 2, 2)$.

We see, first of all, that even using the Extensive form, the two possible ways of using it give different answers. (i) gives rise to a policy $(2, -, -)$ (ii) gives rise to a policy $(2, -, -)$ or $(1, 1, 1)$ depending on which of a group of equivalent alternatives chosen at the second stage. The use of the Normal form gives yet another pair of different policies. If one takes as a (reasonable) postulate, that the Extensive and Normal forms should be equivalent then the Hurwicz α criterion is invalid[1], as a producer of residual values in sequential problems. It is also invalid because equivalent choices at a later stage do not produce equivalent choices at earlier stages if the approach in (ii) is adopted.

Considerations of this kind are extremely important to a computable theory of value in sequential selection processes. Traditional 'decision' theory deals with acts—states of nature—and values of taking an act when the true state is given. Very little of this theory is concerned with the derivation of these values. Such values do not only have to maintain the ordering of the consequences, they have to be sufficient variables on which to base the first stage choice. Where the consequences of an act are comparable with the act itself, as with von Neumann's theory, we can determine appropriate valuation procedures. Let us see what such a postulate would add to a type of theory such as Shackle's (61) theory of choice in as far as value theory is concerned.

Such a theory postulates that four functions exist which describe choice, viz:

(i) $\quad\quad\quad \rho(\theta)$: degree of surprise in outcome θ
(ii) $\quad\quad\quad v(\theta)$: value of outcome θ
(iii) $\quad\quad \phi(\rho, v)$: stimulus function
(iv) $\rho(\phi_{max}, \phi_{min})$: choice function.

For each act, ϕ_{max}, ϕ_{min} are sufficient to determine choice according to function χ. Now if acts and consequences all belong to the same set, χ is a value function itself. Let any particular consequence θ be subjected to the same choice procedure as any act. We do this by identifying each consequence θ with a constant act as follows.

$$\theta \equiv [\theta, \theta, \theta \ldots \theta]. \tag{60}$$

* Except for the case when $\alpha = 0$. This gives us the Wald maximin criterion.

Since $\rho(\theta|\theta) = 0$, we obtain the following value of θ.

$$v(\theta) \equiv \chi(\phi(0, v(\theta)), \phi(0, v(\theta))) \text{ for all } \theta.$$

Thus we obtain the identity

$$x \equiv \chi(\phi(0, x), \phi(0, x)) \tag{61}$$

for all possible values of x.

(61) does not contain the full force of our proposition, and it is possible that much more significant results could be obtained. The important point is that the representation of a consequence as a constant act has useful implications for value theory. Certain forms of ϕ can now be ruled out, and feasible functions explored.

Having said a great deal about value and its part in sequential and non-sequential choice problems, we shall now turn to theories of uncertainty as they are relevant to the problems of choice and decision. Value and uncertainty are complementary concepts in decision theory.

2.3 Uncertainty[1]

2.3.1 Uncertainty as a Measure of Propositional Truth

In Section (2.2) we discussed value as an order preserving homomorphism of choice over a set of elements χ. This is a purely subjective component of choice behaviour and must stem simply from observing choice. It must depend on the person himself. Now the consequence of an act will depend on which proposition, h, in a specific set, H, is true. Thus $\theta(a, s)$ is the consequence of act a when the proposition 'the state of nature is s' is true. Uncertainty theory is concerned with the determination of a real measure on the propositional set H which, together with a value measure, is a determinant of choice.

Uncertainty theories divide into so called 'objective' and 'subjective'. There is also a theory concerned with 'subjectively derived objective' measures. Subjective measures are derivable from observation of choice[2], whereas objective measures are derived, once the basic data is given, by specific procedures, independent of the problem faced. In both cases the measures have significant value

---

[1] We now use 'uncertainty' in the sense of non-certainty.

[2] Values are wholly dependent on choice behaviour.

only when they are not dependent on the consequences and their values which ensue from the truth or falsity of the proposition. This is one reason why objective measures play a big role in decision theory.

Although, when the appropriate data is given, objective measures are determined outside the context of the problem faced, we shall argue that their use in problem solving requires certain implications concerning choice. All such theories are, therefore, essentially subjective in origin.

Let us first of all consider some of the more common treatments of uncertainty.

2.3.2 Some Measures of Uncertainty

A. OBJECTIVE MEASURES

I. Von Mises Frequency Probability

This is the traditional approach to measuring uncertainty in a proposition h. If condition c is observed to occur n times, and proposition h is observed to be true m of these times and false on the other $(n-m)$ occasions, then the 'probability' that h will be true when c is known to be true is defined by

$$p(h \mid c, m, n) = m/n \tag{1}$$

See von Mises (69) for a coverage of this approach.

II. Carnap's Logical Probability

Each one of a set of N individuals is known to have one of a set of Q properties. A Q property may be a p dimensional vector, whose components are dichotomous, giving 2^p possible Q properties. The state of the individuals is determined when the Q property for each individual is known. Two states are equivalent if they differ only in a permutation of their Q properties. Thus $(1, 0, 1, 1) \equiv (1, 1, 1, 0) \equiv (0, 1, 1, 1) \equiv (1, 1, 0, 1)$. A complete set of equivalent states determines a Q structure. If there are k Q properties in all, there are $\begin{bmatrix} N+k-1 \\ N \end{bmatrix}$ Q structures. The logical probability measure of a Q structure is set equal to $\begin{bmatrix} N+k-1 \\ N \end{bmatrix}^{-1}$ The logical probability measure of a state in a given structure is equal to the logical probability measure of the Q structure divided by the number of states in the structure. A sentence is a proposition about the states of the individuals. The probability measure of a sentence is the sum of all probability measures of states consistent with the sentence. A hypothesis, h, is a sentence and hence has a measure. Evidence, e, also has a corresponding sentence and hence has a measure. The probability (or 'credibility' as Tintner (66) calls it) of h when e is known is given by

$$c(h|e) = m(h \cap e)/m(e) \tag{2}$$

where $h \cap e$ is the conjunction of h and e.

Let us consider an example. If we are producing a product over two years which will either be successful or unsuccessful in each year, we have two individuals, a one dimensional dichotomous attribute vector, and hence $N = 2, k = 2$. The number of Q structures is 3 and, if u, s denote unsuccessful and successful respectively, we have the following credibility measures.

$$c(u, u) = 1/3 \, ; \, c(s, s) = 1/3 \, ; \, c(u, s) = 1/6 \, ; \, c(s, u) = 1/6 \tag{3}$$

If, on the other hand, the product had been successful in the first year, and we wish to find the corresponding credibilities in the next two years, we get $N = 3, k = 2$, the number of Q structures is 4 and

$$c(u, u|s) = c(u, s|s) = c(s, u|s) = 1/6 \, ; \, c(s, s|s) = 1/2 \tag{4}$$

III. Classical Logical Probability

This would follow as for Carnap, but all states would be given the same probability measure. Thus (3) and (4) above would become, respectively,

$$c(u, u) = c(s, s) = c(u, s) = c(s, u) = 1/4 \tag{5}$$

$$c(u, u|s) = c(s, s|s) = c(s, u|s) = c(u, s|s) = 1/4 \tag{6}$$

B. SUBJECTIVE MEASURES

I. Savage's Theory of Subjective Probability

Savage's 'subjective probability' is derivable from observing choice and is intimately connected with his theory of value as expounded in Section (2.2.2.2).

There are two forms of probability he considers. These are,

55

qualitative probability (as determined by axioms (1)–(5)) and quantitative probability (as determined by axioms (1)–(5) and (6) or (6^1)).

A qualitative probability relation $\leqslant \cdot$ between pairs of propositions in a set H, satisfies the following axioms.

Axiom 1.

$\leqslant \cdot$ is a simple ordering (satisfies properties (i), (vi), of Section(2.1)).

Axiom 2.

$(h_1 \leqslant \cdot h_2) \rightleftarrows ((h_1 \cup h) \leqslant \cdot (h_2 \cup h))$ provided $h_1 \cap h = h_2 \cap h = \phi$

Axiom 3.

$\phi \leqslant \cdot h \, \varepsilon \, H \, ; \, \phi < \cdot H$

ϕ is the empty proposition;

$< \cdot$ is defined as: $h < \cdot k$ if $h \leqslant \cdot k, k \nleqslant \cdot h.$

$(\geqslant \cdot \equiv \leqslant \cdot')$

To determine whether $h_1 \leqslant \cdot h_2$ we consider two acts as follows:

$a_1 \equiv f$ if h_1 is true, $a_1 \equiv g$ if $\sim h_1$ is true.

$a_2 \equiv f$ if h_2 is true, $a_2 \equiv g$ if $\sim h_2$ is true.

If, for all pairs (f, g) with $f > g$,[1] a_2 is chosen rather than a_1, then $h_1 \leqslant \cdot h_2$.

As we shall see later on, qualitative probability, as with other qualitative relations, has limited use in decision-making. It does have some, and contributes towards partial decidability.

$\geqslant \cdot$ is an ordering relation among propositions, independent of the consequence of the truth or falsity of the propositions. The determination of an order preserving homomorphism of $\leqslant \cdot$ over the real line is then no different to that of value determination. Savage's postulates (6) or (6^1), which allow fine subdivision and comparability of the proposition set H, effectively allow us to construct a quantitative probability measure which preserves the

---

[1] '$f > g$' means 'consequence f is chosen rather than consequence g, and they are not equivalent consequences.'

ordering of $\leqslant \cdot$ and satisfies the usual classical probability laws given below.

$$p(h_1 \cup h_2) = p(h_1) + p(h_2) \text{ if } h_1 \cap h_2 = \phi^1 \qquad (7)^1$$

$$\text{If } H = U_i(h_i), h_i \cap h_j = \phi \ (i \neq j), \text{ then } \sum_i p_i(h_i) = 1. \qquad (8)$$

$$p(h) \geqslant 0 \qquad (9)$$

$$p(\phi) = 0 \qquad (10)$$

It is essential to note that the multiplicative law for independent events is not part of Savage's theory. For this reason the theory, as presented here, has no validity for sequential problems unless the multiplicative law is established separately.

The proposition set H will determine whether we can say anything stronger than $h_1 \geqslant \cdot h_2$ or $h_2 \geqslant \cdot h_1$. If $H \equiv (h, \sim h)$ then, unless $h \geqslant \cdot \sim h$ and $\sim h \geqslant \cdot h$, we can determine neither $p(h)$ nor $p(\sim h)$ within H alone. One way of doing this is to consider the space $H^+ \equiv H \times K$, in which K is a finely divisible space. If we can assume that if $h^+ = h \times K$, then $p_{H+}(h^+) = p_H(h)$, then, since H^+ is now finely divisible, we can determine $p_H(h)$ by finding a $k \, \varepsilon \, K$ such that $h \times K = \cdot H \times k$. In this event, $p_H(h) = p_K(k).^2$

II. Theories of Marschak and Raiffa

Marschak (48) considers problems in which the probabilities, $p_{ij}(a)$, that the outcome will be θ_i when the state is j, are known. Subject to the validity of four axioms of choice, choice is shown to be consistent with a valuation based on values u_i for θ_i and 'probabilities' π_j for states j, which takes the following form.

$$v(a) = \sum_i \sum_j u_i \, \pi_j p_{ij}(a). \qquad (11)$$

The π_j's are the *a priori* subjective probabilities for states j.

Varying the acts varies the $p_{ij}(a)$'s.

Raiffa and Schlaiffer (57) consider a similar situation in which the values of outcomes are assumed. On the basis of three axioms, choice is shown to be in accordance with an act valuation as follows.

$$v(a) = \sum_j \pi_j u_j(a) \qquad (12)$$

---

[1] ϕ will refer to the empty set unless otherwise defined.
[2] The suffix H^+, H, K denotes the spaces within which the probabilities are measured.

The π_j's are the *a priori* subjective probabilities of states j.
Varying the acts varies the consequence values.

III. *Luce's Subjective Probability for Probabilistic Choice*

In Chapter Two, Section 2.2.2, we discussed Luce's theory of probabilistic choice, in which we did not require that choice should be consistent. Given $Q \equiv \begin{bmatrix} a \\ b \end{bmatrix}$ sometimes a would be chosen, and sometimes b would be chosen. As Luce assumes in his theory, the same point applies to one's belief in the truth or falsity of propositions. Luce introduces a quantity $Q(\alpha, \beta)$ which is the probability that if a person were asked to choose between propositions α, β he would choose α. This choice behaviour must be independent of the consequences of α, $\sim\alpha$, β, $\sim\beta$.

Luce's subjective probability, $\phi(\alpha)$, is then simply an order preserving homomorphism of the relation $\geqslant$, defined as follows.

$$(\alpha \geqslant \beta) \rightleftarrows (Q(\alpha, \beta) \geqslant \tfrac{1}{2}) \qquad (13)$$

This is similar to Debreu's approach to value. The $Q(\alpha, \beta)$'s have to satisfy certain properties before $\geqslant$ becomes a simple ordering.

Luce gives no such pre-requisites, but a straight transposition of Debreu's axioms from acts to propositions would suffice.

Although ϕ preserves the order of H, it lacks one of the usual properties of probability measures. If α does not intersect β, we can have $\phi(\alpha \cup \beta) \neq \phi(\alpha) + \phi(\beta)$.

IV. *Shackle's Degree of Surprise*

So far we have been using the conventional approach to uncertainty involving probabilities of several kinds which, however, bear a great deal of similarity to each other. Shackle (61) feels that probability approaches are unrealistic as co-determinants of choice, and introduces the idea of 'degrees of surprise'. The manner in which choice is based on these is given in Section (2.2.2.2). I shall quote his postulates for degrees of surprise, since they constitute the major deviation from probabilistic theories evident in the literature.

Axiom 1. An individual's degree of belief in a hypothesis can be thought of as consisting in a degree of potential surprise associated with the hypothesis and in another degree associated with its contradictory.

Axiom 2. Degrees of surprise can be zero or non-zero, with maximum $\bar{\rho}$, signifying complete disbelief.

Axiom 3. Equality between the respective degrees of belief felt by an individual in two hypotheses will require, for its expression in terms of potential surprise, two statements, viz: that some given degree of potential surprise is attached to both hypotheses and that some given degree is attached to the contradictories of both.

Axiom 4. The degree of potential surprise associated with any hypothesis will be the least degree among all those appropriate to different mutually exclusive sets of hypotheses (each considered as a whole) whose truth appears to imply the truth of the first hypothesis.

In other words if $h = U_i h_i$, $h_i \cap h_j = \phi$, $i \neq j$, then

$$\rho(h) = \min_i[\rho(h_i)]. \tag{14}$$

Axiom 5. All members of an exhaustive set of rival hypotheses can carry surprise.

Axiom 6. When h is any hypothesis the degree of potential surprise attached to the contradictory of h is equal to the smallest degree attached to any rival of h.

Axiom 7 (first version).

The degree of potential surprise attached to the joint (simultaneous) truth of two hypotheses is equal to the greater of the respective degrees assigned to the separate hypotheses.

Axiom 7* (second version).

Let ρ_A^B be the degree of potential surprise assigned to hypothesis B when ρ^A is the degree assigned to hypothesis A, and let ρ_0^B be the degree assigned when $\rho^A = 0$. Then

$$\rho_A^B \leqslant \max\left[\rho^A, \rho_0^B\right]. \tag{15}$$

Axiom 7** (final version).

$$\rho_A^B = \max\left[\rho^A, \rho_0^B\right]. \tag{16}$$

Axiom 8. Any hypothesis and its contradictory together constitute exhaustive set of rival hypotheses.

Axiom 9. At least one member of an exhaustive set of rival hypotheses must carry zero potential surprise.

Comments on Shackle's theory can be found in references

(6, 17, 27, 28, 34, 50, 53, 55, 59, 67, 77)[1]. Dunlop (27) indicates that, in his experience, Shackle's theory is correct. One must ask to which part of the theory he is referring to and in what form of interpretation. It is not clear whether the theory is based on observation of choice or on verbal communication. Weaver (71) suggests a relationship between degrees of surprise and objectively given probabilities. Krelle (43) suggests an alternative approach to Weaver's relationship.

The main point arising from such considerations is that this is a radically different treatment to those of probability theorists, at least in respect to the rules of operation with such measures. Unfortunately, as it stands, the theory is incomplete. It does not allow us to determine, on the basis of choice, inequalities among degrees of surprise of alternative hypotheses. Neither does it tell us how to determine quantitative measurements. Verbal statements are of no use in decision theory, since there is no guarantee that they are in accordance with behavioral choice. It is only from observation of choice that we obtain our premises for decision theory. An operational definition of 'degree of surprise' is needed, and then the theory can be given serious attention.

C. SUBJECTIVELY DERIVED OBJECTIVE MEASURES

So far we have dealt with theories in which measures of uncertainty are either formally derived from specified data, or are imputed by observing choice in a given class of problems. It is perhaps not an unreasonable pre-requisite that objective and subjective measures should be correlated to some extent. Good (31) takes the view that there are such things as objective probabilities which have an absolute nature (i.e. have some 'real' interpretation[2]), but whose values are not known precisely and are, therefore, subject to subjective assessment. This is different to the theories of Savage and such people as Marschak, in which probabilities are co-determinants of choice, and are not objective probabilities. Good would accept answers to questions of the form 'what is the probability of proposition h?'. Such probabilities may not be the same as those determined

---

[1] Perhaps the most serious disadvantage is that, given gambles $g_1 \equiv E\pounds 1\,\tilde{E}$, $g_2 \equiv F\pounds 1\,\tilde{F}$. where $E = (H, H)$, $F = ((H, T) \cup (T, H))$ in the toss of two coins. such a person may prefer g_2 to g_1 and yet give zero surprise to E, $\tilde{E}$ and F, $\tilde{F}$. This contradicts the validity of surprise as a joint determinant of choice.

[2] The reader is advised to refer to Chapter (1) concerning the 'meaning' of a concept.

by Savage using experimental problematic situations. There is no guarantee that such probabilities would satisfy any probability laws.

There is, perhaps, a fine dividing line between determining measures either by observing choice in problematic situations or by asking people to give verbal estimates. Both are subject to unidentifiable mental processes, both, therefore, being subject to 'errors' in several ways. The real dividing line is one of inference in relation to choice. Savage says choice is consistent with a valuation procedure using certain probability measures. Good says nothing about actual choice.[1]

Although we have discussed uncertainty in terms of 'probability' or 'degrees of surprise', many terms have been used in practice when talking about propositional truth measures. Some of these are: confidence (Reichenbach (58)), credibility (Good (31)), confirmation (Carnap (16), Tintner (66)), surprise (as listed in references 6, 17, 27, 28, 34, 50, 53, 55, 59, 67, 77) acceptance (Shackle (62)), belief (Good (31)).

Having briefly discussed some of the theories of uncertainty measures put forward, let us now try to ascertain further the relevance of the objective theories to theories of choice. Some of this relevance has been discussed, in discussing theories of uncertainty, being incidental to an understanding of the theories. There are, however, certain general points which need discussing.

2.3.3 Objective Uncertainty Measures and Choice

Objective measures of uncertainty are simply abstract measures on the proposition space H, divorced, entirely from choice, and dependent on a specified knowledge content, K. Subjective measures are measures on the proposition space H, determined by observing choice, and are co-determinants, together with value, of choice. The problem is, therefore, raised as to the relevance of objective measures to choice. We have to question their appropriateness for choice purposes, as far as they represent the content of K relevant to propositions in H. We need also to question the manner in which such measures co-determine choice, together with value, in specific problematic situations.

---

[1] Although he mentions the use of expected values as a decision procedure, this is a normative recommendation and is not derived, in his theory, from actual choice characteristics.

Objective probability measures are constructed with a twofold purpose in mind. The first is the condensation of a body of knowledge K to a form which is more readily comprehensible. The second is the use of the probability calculus in being able to determine uncertainty measures of compound events from the uncertainty measures of the component events. The probability calculus plays a parallel role to that of value structures. The first point is exemplified if a mass of data exists which has some bearing on the truth or falsity of propositions concerning the demand for a particular product. The data, itself, may be incomprehensible, but when it is reduced to probabilistic statements it may be more readily understood. This purpose is not in itself sufficient of course. Thus Carnap's measure will achieve this purpose, but is generally unacceptable. Such measures must have certain properties in certain problematic situations, as we shall discuss below.

It is implicit in the use of such measures, that any two propositions with the same measure will be treated as equivalent in any problems depending on their truth or falsity. Thus if we toss two coins, if Carnap's measure is to be used, then the events (H, H), $((H,T) \cup (T, H))$, (T, T) are all equivalent in problematic situations since these have the same credibilities. It is because this equivalence is not generally acceptable that Carnap's measure is not generally valid as a co-determinant of choice.

An example in which even classical probabilities are insufficient for choice can arise when events with the same probabilities but different physical specifications are used. In one case E_1 occurs if, say, a black card is drawn from a shuffled pack of cards. In another case E_2 occurs if 25 or less black cards occur in 51 cuts of the pack (with reshuffling). Both have the same classical probabilities of $\frac{1}{2}$, and yet it has been found that the events are not equivalent in problematic situations in general. Thus probability alone may not be sufficient for describing events for the purpose of choice.[1]

A stronger property is required than that of equivalence. Probabilities must also have certain order preserving characteristics of choice. Let $p(h_1) > p(h_2)$. Construct the problem, Q, as follows:

$$Q \equiv \begin{bmatrix} a_1 \\ a_2 \end{bmatrix} \tag{17}$$

[1] Of course many may accept that not to treat the two cases as equivalent is unreasonable, but we are concerned, at this stage with actual rather than Normative behaviour.

where

$$a_1 = \begin{array}{l} 1 \text{ if } h_1 \text{ is true} \\ 0 \text{ if } \sim h_1 \text{ is true} \end{array}$$

$$a_2 = \begin{array}{l} 1 \text{ if } h_2 \text{ is true} \\ 0 \text{ if } \sim h_1 \text{ is true} \end{array}$$

We would expect $a_1 > a_2$.

This is, in fact, part of Savage's theory and we see that we are converging on Savage's subjective approach. The difference is that a procedure is proposed for computing probabilities directly from (K, H) and its validity is tested by requiring that certain choice behaviour obtains, whereas Savage deduces his probability measures from observing whether or not certain axioms of choice are valid.

If it were not for the severe difficulties in determining subjective probabilities, and the wish to reduce these, we would not bother with objective probabilities. However, in the case where background knowledge is complex, it may not only be easier to use a formal approach, but, because of mental limitations, it may be thought more 'reliable' to do so[1].

If we use probabilities of any kind, we need to be able to interpret the statement 'the probability, that the proposition h is true, is p'. This is carried out by identifying some event whose probability (in the language used) is p. Thus if we are told 'the Carnap probability of h being true is $\frac{1}{4}$' then we may identify h with the proposition 'the outcome of tossing two coins is (H, H)'. If it is frequency probability language we identify h with any 1 in any sequence of 1's and 0's in which there are a proportion of $\frac{1}{3}$ 1's.

The manner in which choice depends on a particular form of probability measure may be tackled on similar lines to von Neumann's approach. This is achieved by setting up experimental problematic situations in which probability mixtures (in any specific language) of outcomes constitute the set of alternatives, M.

Whether analyses are carried out in terms of subjective or objective measures of uncertainty must depend on the appropriateness of both. In some cases a mixture of subjective and objective probabilities may be used, providing the appropriate assumptions are made.

---

[1] We do not define reliability, but simply point out that a person's assessment of probabilities may be, intuitively, very wide of the mark for several reasons. Helmer (80) discusses ways of measuring reliability and accuracy in probability estimation.

Thus, in sampling problems, subjective assessments of population parameter distributions may be used, and objective probability measures for observations (for a given population parameter) combined with the subjective parameter measures to estimate the effect of sampling strategies. We are thus combining both purposes of probability measures, viz. the representation and compounding of information. Marschak's (48) work is specifically oriented this way. The purely subjective aspect is of use when it is impossible to determine or use the precise knowledge and content of, say, an expert.[1] In using it we, however, must remember its precise meaning in the context of Savage's theory. When the knowledge content is formally presentable, then objective probabilities become more appropriate. We will now pay some attention to the problem of knowledge in as much as its influences the theories of uncertainty discussed previously.

2.3.4 Uncertainty Measures, Knowledge and Propositional Forms [2]

Whether we use objective measures or subjective measures, subjectivity must enter into their derivation. Such measures must depend on the knowledge content, K, of the individual.

Theories such as those of Savage and Shackle make no reference to the knowledge content, K, of the individual concerning the events defined in the theories. As a consequence, if we tried to determine the corresponding measures for two different people we would expect that they would not agree because the knowledge contents would be different. This sort of consideration is likely to be to the disadvantage of advocates of Good's theory, in general. In special cases, in which K can be assumed to be reasonably uniform over the people concerned, then we may get close agreements. However, this raises the problem of determining knowledge content.

It has been stated earlier on that one purpose of determination of probabilities, subjective or otherwise, is that the 'content' of a body of knowledge held by one person can be communicated, in a

---

[1] We do, however, run into difficulties since such a measure is pertinent to the expert's characteristics and the decision-maker, even if he had the same information, may have different subjective probabilities.

[2] The words 'knowledge' and 'information' (used later on) are interchangeable. In the literature it is usual to refer to knowledge, if we are concerned with the subjective aspects of theories of choice. Information theory has been developed as a separate area of study.

comprehensible form, to another. In the case of subjective measures we make no analysis of knowledge content at all. This raises the important problem of why one set of probabilities is to be accepted as more reasonable than another set. This may be purely an act of faith, based on the so called 'expertise' of the person whose probabilities are being measured. Precisely how one defines 'expertise' is another matter altogether. Helmer (37) discusses this problem. Some quite interesting problems arise here.

We may have some degree of confidence in the expert's appraisal and this introduces uncertainty at a different level. If we have two people giving appraisals, and these differ, we have three possibilities, any one of the appraisals being 'right' or neither being 'right'. Helmer (37) devises measures of reliability and accuracy for such probability measures which enable a partial resolution of these problems.

Little or nothing has been done in the area of a theory of knowledge and its relevance to choice. The difficulties lie in determining what it is a person knows. Churchman (20) says that 'we only know what a person knows by observing choice and knowing choice procedure'. This leads us in the same direction of Savage, who believes no theories of uncertainty are meaningful outside the context of choice. The difficulties also lie partly in the fact that what a person knows may only be what he believes. Thus, this knowledge content, K, contains a set of propositions assumed by him to be true, but which may or may not be true. They may have been incorrectly deduced from other propositions. They may have been communicated to him as propositions 'likely' to be true and accepted as being true. The process of communication can also result in errors.

The proposition space, H, is subjective in nature and complicates the problem of deriving probability measures.

If we are dealing with frequency probability measures, then the selection of conditions, c, is subjectively influenced by the person concerned. Thus, if we are trying to estimate the probability that the demand for a product will be at a certain level, there are an enormous number of variables which could conceivably be included in the condition vector c. We could include economic parameters (various), population size, election imminence, and product characteristics. Assuming we can identify the levels of each at each point in time the measures obtained for each form of c must inevitably differ. Thus, although we may have complete knowledge of every possible condition factor at each time, we would still have to select which factors were to go into the model. Some would be 'felt' to be important, and others 'felt' not to be important.

An attempt to bring in too many factors would run into the difficulty that there would then be conditions which had never been realized in the past. It is possible, with enough factors, to find a condition format in which no two conditions have ever been identical. This would result in probabilities of doubtful value.

Such difficulties arise using Carnap's measure, since the choice of the vectorial form describing the condition of an individual can influence the values of the logical probability of a proposition. Consider the problem of determining Carnap measures for propositions about the demand for a new product for which we have no prior evidence. Suppose the demand is classified as good or bad, and that two dichotomous attribute factors are used referring to good or bad economic conditions and to good or bad product competition. Assume that neither of the levels are known in advance. We then have 8 Q properties in all. If we are considering a two year span, we have 36 Q structures out of which 10 will correspond to the hypothesis 'the product demand will be bad in both years', giving rise to a logical probability of 5/18 for this hypothesis. If, however, we only consider one factor (e.g. economic conditions), we would obtain a logical probability of 3/10 for our hypothesis. Since we are postulating no prior evidence, then we have no reason to believe that economic conditions or product competition influence demand, and we have no way of 'deciding' between the two probability figures. To avoid misinforming the problem solver, we would have to provide him with the bases of derivation, and then he provides the subjective assessment of the acceptability or otherwise of the bases.

With Shackle's theory we encounter a similar form of difficulty in relation to the proposition set H. This arises because of the form of mental comparison which is carried out when we say 'I shall be surprised if proposition h is true'. Consider a sequence of ten tosses of a coin. Shackle would not be surprised if ten heads turned up, providing he was considering each possible proposition corresponding to a specific realization of the ten tosses. In effect he is considering them pairwise, and, as a consequence of his axioms, each realization would have zero degree of surprise. However, another person might be extremely surprised if ten heads did come up. In this case, it is likely that he is considering only two propositions, viz. $h \equiv$ 'ten heads will turn up' and $\sim h \equiv$ 'ten heads will not turn up'. Weaver (71) emphasizes this possibility in trying to construct a one-one homomorphism between Shackle's degree of surprise, $\rho(s)$ and classical probability measures, $\hat{p}(s)$. The relationship he suggests is dependent on the set of mutually exhaustive and exclusive propositions conceived, and is given below.

$$\rho(s) = (\sum_s \hat{p}_s^2)/\hat{p}_s \tag{18}$$

$$\hat{p}(s) = 1/(\sum_s \rho(s)^{-2})\rho(s). \tag{19}$$

In the example we have quoted, the degree of surprise would be 1 and $(1+(2^n-1)^2)/2^n$ respectively, the latter being much larger than the former. Weaver's lower limit is not zero, but this does not matter within the context of the point at issue.

When we get to the stage of using statistical models for determining relationships between variables, we shall face much the same problem. Propositions will then take the form

$$z = \phi(x)+\varepsilon \tag{20}$$

where ϕ is some function to be chosen, and ε is to be specified as to its probabilistic properties.[1]

With no prior evidence we have no reason to choose one form rather than another. Since, by introducing enough factors into x, ϕ can be chosen to make $\varepsilon(z, x) \equiv 0$, we could always get a non-probabilistic propositional form for z given x. This sort of approach violates the principle of Occam's Razor, and is likely to be unacceptable to most scientists.

The problem of subjectivity in objective measures, can be resolved by treating propositions used in building the models from a subjective uncertainty point of view. This is the point of Baysian analysis, which we shall discuss later. It is embodied in Marschak's work, and in Good's work, who constructs the basis of a theory in which he allows for a body of beliefs on which the models are to be built. These approaches are important, for our formal model building is limited to being a mixture of subjective and objective elements. We cannot get away from it.

---

[1] On the basis of this we might inform the decision-maker that his product demand will be 1,000 units with a probability of ·9 of the deviation being no more than 100 units. All we are saying is that relation 20 has a certain predictive power; we say nothing about any intrinsic probability of demand.

2.3.5 *Principle of Insufficient Reason*

A principle which is quite often referred to in determining initial probability measures, objectively, over a set of propositions, is the 'Principle of Insufficient Reason' (Savage (60)). This principle purports to overcome the subjective elements in probability derivations by determining them on the basis that the 'information' available, reflecting on the truth or falsity of a set of propositions, is 'symmetrical' in some sense. It is difficult, however, to define what 'symmetry' is. Savage states: 'The "Principle of Insufficient Reason" depends upon the "symmetry of information" and the "consistency of reasoning". It is not always obvious what the symmetry of information is in a situation in which one wishes to invoke the principle of insufficient reason.'

Classical logical probability measures depend on this symmetry. The determination of a division of the set H, in which each proposition has the same measure of probability, operates on this principle. Failure to answer the problems of 'symmetry of information and reasoning' must lead us back to our subjective basis for determining initial probability measures. A definition of symmetry, based on choice, is given by Milnor (51), in the form of an axiom of choice. It reduces to the condition that all events (or propositions) are interchangeable in any problematic situation. This was discussed in Section (2.2.2). Thus we would say that there is symmetry of information about $h_1 \equiv$ 'the demand will be between 10 and 20 units' and about $h_2 \equiv$ 'the price will be between 1 and 6 units' if all problems based on the truth or falsity of either h_1 or h_2 result in the same choice if h_1 and h_2 are interchanged.

The fact that the principle of insufficient reason apparently applies to such things as the roulette wheel, dice games, and coin tossing, arises, not from the absence of reason, but from the very use of reason. As Reichenbach (58) points out, the near equi-probabilities of certain propositions can be 'deduced' or 'decided'.

Let us beg, for the moment, the question of determinism, and assume, for the roulette wheel, that the angular position at which the ball will rest is completely determined by some known initial condition, c. Let us assume that if θ is the angular position of rest, and c contains an initial angular speed, initial position of the ball, and other factors. Thus we have the following deterministic equation.

$$\theta = \phi(c). \tag{21}$$

Let us assume that c has some known probability density function, $g(c)$, continuous in c. Thus we postulate the existence of probabilities, but make no statement about their form.

Let the roulette wheel be broken up into $2n$ regions of alternately red and black.[1] Assume that red occurs in the region $(\theta_{2r-1}, \theta_{2r})$, and black in the region $(\theta_{2r}, \theta_{2r+1})$, $r = 1, 2, \ldots n(2n+1 = 1 \bmod(2n))$.

Let $p_n(R)$, $p_n(B)$ be the respective probabilities of red and black (assuming classical probability calculus). Let $f(\theta, g)$ be the density function of θ, given g. We then determine the following expressions:

$$p_n(R) = \sum_{r=1}^{n} \int_{\theta_{2r-1}}^{\theta_{2r}} f(\theta, g)d\theta \qquad (22)$$

$$p_n(B) = \sum_{r=1}^{n} \int_{\theta_{2r}}^{\theta_{2r+1}} f(\theta, g)d\theta \qquad (23)$$

Now, using the Mean Value theorem for integrals, and assuming the analytic properties required for, f, we get

$$p_n(R) = \sum_{r=1}^{n} (\theta_{2r} - \theta_{2r-1})f(\alpha_r, g) \qquad (24)$$

$$p_n(B) = \sum_{r=1}^{n} (\theta_{2r+1} - \theta_{2r})f(\beta_r, g) \qquad (25)$$

where

$$\theta_{2r-1} \leqslant \alpha_r \leqslant \theta_{2r} \qquad (26)$$

$$\theta_{2r} \leqslant \beta_r \leqslant \theta_{2r+1} \qquad (27)$$

Letting $\theta_{2r} - \theta_{2r-1} = \theta_{2r+1} - \theta_{2r} = \Delta = \pi/n$ we see that

$$p_n(R) - p_n(B) = (\pi/n) \sum_{r=1}^{n} (f(\alpha_r, g) - f(\beta_r, g)) \qquad (28)$$

Again, using the Mean Value theorem, we obtain

$$f(\alpha_r, g) - f(\beta_r, g) = (\alpha_r - \beta_r)\frac{\partial f}{\partial \theta}(\gamma_r, g) \qquad (29)$$

where

$$\alpha_r \leqslant \gamma_r \leqslant \beta_r. \qquad (30)$$

---

[1] This is thus a simplified roulette wheel.

Now, if we restrict ourselves to density functions in which $\left|\dfrac{\partial f}{\partial \theta}\right|$ is uniformly bounded by some number K, since $|\alpha_r - \beta_r| \leqslant 2\pi/n$, we obtain

$$|p_n(R) - p_n(B)| \leqslant (2\pi^2/n^2)K \qquad (31)$$

The actual bound is likely to be much lower, but the point is clear that the near equiprobability of R and B depends on n. The same applies to coin tossing, providing we ensure enough variation in the number of spins at the time when the coin comes in contact with the ground.

The point of this analysis is that if n is large enough we need not worry too much about $g(c)$, although there will be a residual small bias which we do not know of course. When n is small, accepting the validity of (21) (or its counterpart in other situations), we can control the values of $p_n(R)$ and $p_n(B)$ quite easily. Thus, if $n = 1$, in the roulette problem we could reasonably determine the region of c values which would almost certainly give rise to R. In coin tossing we could, by making the number of spins small, almost guarantee to get heads or tails each time by controlling propulsion conditions initially.

Equiprobabilities could also be achieved also by controlling $g(c)$. Again, however, the equiprobabilities are deducible. Thus for $n = 1$, in the roulette problem, we could arrange for a small impulse to be applied each time, beginning with the ball in the middle of the red and black regions alternately. This situation would be deterministic for most positions of the ball.

The points made above may have an important bearing on logical probability in certain areas in which there exists relationships of the kind discussed. In the production of products which are either defective or non-defective, objective, probabilities might conceivably be independent of the probabilistic form of some determining variables. Thus certain settings may be fixed, apart from small variations, and then the small variations might then be a determinant of output quality.

Given the existence of laws, such as (21), some probabilities may be, within the limits of measurability, almost predetermined in a logical manner.

We shall now discuss the problem of determinism versus limiting uncertainty. This is an essential element in any pragmatic theory of decision, particularly since, according to some interpretations, 'better' decision-making involves the quest for certainty of outcome.

2.3.6 Principle of Determinism versus Principle of Uncertainty

There are two opposed schools of thought concerning the possibility of reducing all uncertainty in propositional truths to zero. One school, of which Einstein is a member, believe that ultimately it will be possible to determine enough knowledge about things in general for all propositions, which mean anything, to be classifiable as true or false. Thus, for example, it will be possible to determine the form of conditions, c, in the frequency probability approach, such that $p(h|c)$ takes only the value 0 or 1. Such a point of view is the guiding aim of scientists whose prime purpose is searching for truth in various areas. They are not, in general, people who evaluate their uncertainties against an economical background.

The other school, among whose members is Barankin (4), believes that there is an ultimate level of uncertainty in empirical propositions, beyond which we will not be able to go. Thus a barrier exists not only because of practical limitations (because our tools will always be inadequate for the purpose, and will, in any case, disturb the system in their very use), but because, as Barankin believes, nature is one huge stochastic process. He says that even physical elements, such as electrons, can, in a sense, be said to 'choose'. They are thus unpredictable, and no laws exist which would ever, however we constructed the conditional propositions, predict such behaviour. The growth of quantum mechanics and the importance of Heisenberg's Principle of Uncertainty[1] arise from this belief. Any attempt to measure anything is an operation on the system which 'changes' the system's behaviour somehow, and we never know how. Only by aggregating can we validly talk about measures. We cannot hope to measure the velocity of a single electron because our tools are too crude. We have to consider an ensemble of such electrons, and talk about 'average' quantities.

These two points of view pose difficult problems for our theories when we come to use them. Quite obviously, even if the world is essentially deterministic, we do not search out its determinism at any cost. Information is costly. Bearing this in mind, limiting uncertainty

---

[1] This effectively states that we cannot, at the same time, measure accurately two variables since the process of measuring one disturbs the system for measuring the other.

may be preferable to certainty at a cost. In any case it is by no means obvious that a person would want determinism even if it were not costly. This poses a difficult problem when we consider the value of information. It is also important to realize that, if everyone had the certainty requirement, some people who, under uncertainty, had some chance of success, may not then have any at all. Uncertainty may be desirable in life in general. It is argued that this is true enough when one is not involved in such serious problems as controlling a business, but that this could never be accepted as a dogma in business. This merits serious consideration for, after all, business is controlled by humans.

One serious drawback of the determinism principle is that, unless new conditions are prevented from arising, we can never know the relevance of a new condition to the truth of a proposition. Hence if we accept the fact that new conditions will, by virtue of a person's 'free will', always arise, determinism must fail. If, of course, no such free will is allowed, each person is forced to act in accordance with some theory of choice and the process is ultimately closed[1], we may have determinism. However, even without 'free will', the generation of new conditions may continue.[2]

Taking the line of the uncertainty theorists, we do meet the paradox that if we cannot determine the truth value of a proposition, then it follows that we cannot determine the truth value of a frequency probabilistic proposition unless it is a logical probability. Therefore, adherents of the principle of uncertainty run into difficulties if they wish to be 'scientific' about their own propositions. Just as we face the possibility of divergence in a deterministic system, we must face it even more so in a stochastic system. For simple closed systems we will have a repetitive effect and the ability to control equipment performance proves this. In complex human situations this will not arise. One thing which illustrates this point is the increasing variety of products and types of leisure to which people are now subject, rendering it difficult to determine a basis for choice, when one has to consider future situations which may arise.

Finally let us note that there is a difference between necessary and unnecessary ignorance. A determinist would say that no ignorance

---

[1] By this we mean that the states into which the system can move belong to a given bounded set out of which the system cannot move.

[2] This has a bearing on values since, as conditions change, values will change, e.g. as more things are available for purchase the value of cash changes.

is necessary, given enough time and manpower to resolve it. There are cases in which ignorance could have been removed. Thus one may take a personnel decision without checking on qualifications, but this information could have been obtained; one may decide to buy a machine without surveying the market for others, but this information is available. On the other hand is it not impossible to find out exactly how much of a product will be sold at a given price? A distinction between the two types of ignorance may be useful. Thus it might be possible to establish the degree of improvements on actual outcomes which would have arisen had more available information been used. We cannot do this when necessary ignorance is involved.

2.3.7 Collective Uncertainty Measures

So far we have been talking of uncertainty measure on individual propositions, h, in a proposition set H. There is a sense in which a person refers to uncertainty 'as a whole' in a process, as distinct from 'individual' propositional uncertainty. 'The outlook is very uncertain' is a common expression.

Collective uncertainty measures may shed no light on the different likelihoods of various consequences for different actions and hence have no use in problem resolution. If the measure is such that these likelihoods can be deduced, then they may serve a useful purpose.

The most common measure of collective uncertainty, in cases where the propositions are of the form '$x = X$',[1] is the 'variance', σ^2. This is defined as follows:

$$\sigma^2 = \sum_X p(X)(X - E(x))^2 \tag{32}$$

where the mean, $E(x)$, is defined by

$$E(x) = \sum_X p(X) X. \tag{33}$$

Corresponding integral forms exist when a density function, $g(X)$ is used instead of discrete probabilities.

In cases where X is Normally distributed, E, σ^2 are sufficient to derive $g(X)$ for any X. It is traditional in much work in statistics (particularly in quality control) to treat E, σ^2 as sufficient, irrespective of whether Normality is valid or not. In such cases, σ^2 is the

---

[1] x is a cardinal variable.

measure of collective uncertainty, which can, together with E, be used to determine individual probabilities.

The second collective measure of uncertainty which plays a significant part in Information and Communication theory (see Khinchin (40)), is that of 'entropy'. This is defined, in the discrete case, as follows.

$$E = -\sum_X p(X)\log_2 (p(X)). \tag{34}$$

A corresponding form for the continuous case exists.

There is no specification of the nature of X. X need not be a cardinal variable, and, in fact, the space may contain any propositional forms. Whereas σ^2 requires H to be a cardinal space, E does not[1]. It is concerned only with the probabilities and not with other measures on H.

Although E plays an extremely important part in measuring information content and rates of communication (information transmission) since it has no bearing on individual propositional truth values, it is of no use in theories of decision. We cannot work back from E to find the individual $p(X)$. People do talk about 'uncertainties in the market', and collective measures are therefore evident in actual problem solving. However, I see no way in which they can be used, formally, for problem solving, other than through the ability to work back to the individual probabilities. In such cases the theories of choice may be framed in terms of such measures of uncertainty.

Theories of choice in terms of E and σ^2 do exist. Lichtenstein (44) considers experiments in which the hypothesis is tested that choice depends only on E and σ^2 in certain problematic situations. Markowitz (47), taking certain axioms of behaviour as valid, and using similar approaches to the axiomatic theories of Savage, Milnor, and von Neumann, deduces that behaviour is characterized by a value function of a specific kind. The axioms are as follows.

Axiom 1. The outcome of an act can be expressed in terms of a cardinal variable r.

Axiom 2. Probabilistic propositions are valid in relating the acts to the outcomes.

---

[1] Thus if $H \equiv (h, \sim h)$ and $h \equiv$ 'the machine is working', then if $p(h) = \cdot5$, $E = 1$, but σ^2 has no meaning for H.

Axiom 3. A function $f(r)$, exists so that $E(=E(r))$ and $F(=E(f(r)))$ are jointly sufficient state variables for describing choice.

The results are that there exist numbers, a, b, c such that choice is in accordance with the value function given below.

$$v(E, F) = a + bE + cF \qquad (35)$$

If we now return to our collective measure of uncertainty σ^2, and if choice is based, say, on E and σ^2 $(= E(r - E^2))$ then

$$u(E, \sigma^2) = u(E(r), E(r - E)^2) = u(E(r), E(r^2) - (E(r))^2) = v(E(r), E(r^2))$$

$$(36)$$

Thus we obtain the following form for $v(E, F)$.

$$v(E, F) = a + bE(r) + cE(r^2) \qquad (37)$$

$$u(E, \sigma^2) = a + bE(r) + c(E(r))^2 + c\sigma^2 \qquad (38)$$

As a consequence of (38), if we assume that, for a given σ^2, larger $E(r)$'s are preferred, then $c \geqslant 0$. Hence for a given $E(r)$, unless $c = 0$, larger variances are also preferred. Thus, if as appears to be the rule rather than the exception, people do prefer less uncertainty for a given level of E, the formulation of choice in terms of a value function, depending on one collective measure of uncertainty, must be suspect.

2.3.8 *Probabilistic Inferences and Probability Chains*

Let us now assume that probability, in some form or other, is a valid form of uncertainty measure, and discuss the forms of probabilistic inference which may arise in some of our formal models later on. The two basic types of inference which are made are probabilistic inferences about observations, and probabilistic inferences about probabilities. Tinter (66) considers three forms of inference, based on these two classifications,

(i) *Direct Inference.* A population, Θ, of propositions (concerning observations) is assumed to be describable probabilistically by some parameter, θ, so that the probability distribution over Θ depends only on θ.

(ii) *Inverse Inference.* A population Θ, of parameters θ, is assumed, and a probability distribution over Θ is inferred from past observations.

(iii) *Predictive Inference.* A population, H, of propositions

75

(concerning observations) is assumed, and a probability distribution over H is inferred directly from past observations.

Quite obviously (iii) may be a combination of (i) and (ii), but need not be. This all depends on whether the second type of inference is thought to be valid. For it to be valid, it is essential that observations are eventually possible which determine the truth values of propositions $\theta \, \varepsilon \, \Theta$. Thus, in batch sampling problems, the proportion of defective items of a batch will eventually be known in cases where 100 per cent inspection is carried out or the customer keeps a record of actual defects himself. In cases where the true value of θ is not knowable then severe conceptual problems can arise. Thus if, in a time-series analysis, we postulate $x(t) = a + bt + \varepsilon$ then a, b are not knowable and yet we carry out many analyses in this form.

The assumption of a population adds great powers to our analytic methods in the cases where the acquisition of information, by taking observation, is an integral part of our problem, since such acquisition may be costly in time, manpower and money. As a starting point of the analysis it can be quite useful. This starting point may come from using objective inverse measures, as given by Carnap, or the corresponding classical case, or it may be subjective. For example, let us consider, very crudely, the set of possible hypotheses for a person's forecasting ability, which we classify as correct or incorrect. We might postulate that, in n attempts to forecast something, he will be right m times, and wrong $(n - m)$ times. We do not know m but we may know that so far, in two forecasting efforts, he has been correct. We may wish to form some probability judgement about the true value of m. We, therefore, postulate that our population contains n elements, m of which correspond to a successful forecast. Carnap's logical probability for m, given two successes[1], is then:

$$c(m \,|\, s, s, n) = m(m - 1) \Big/ \sum_{k=0}^{n-2} (k + 1)(k + 2) \tag{41}$$

When n becomes very large, the corresponding density function $c(\lambda \,|\, s, s)$ is given by

$$c(\lambda \,|\, s, s) = 3\lambda^2 \tag{42}$$

where

---

[1] Carnap's theory is used to show that, notwithstanding its difficulties, it can, formally, still fit within the framework of probabilistic inference.

$\lambda = Lt\ (m/n) = $ limiting long run proportion of successes. (43)

$\quad n \to \infty$

If we use classical probability measures, (41) is replaced by

$$p(m\,|\,s, s, n) = \begin{bmatrix} n - 2 \\ m - 2 \end{bmatrix} 2^{-n+2} \qquad (44)$$

The corresponding density function is much more difficult to obtain.

It may be possible to obtain subjective probability estimates for population parameters, but this is likely to be more suspect than subjective probabilities for single observations. However, if a person behaves as if he attached certain probability levels to population parameters, then this is no different, in principle, to his behaving as if he attached probability levels to certain observations. The reason we say 'more suspect' is that in many instances the person may not think in terms of populations—he may think directly in terms of predictive inference.

However, even though subjectivity may not be acceptable in putting probability figures to each $\theta\ \varepsilon\ \Theta$, considerable subjectivity must exist in the choice of Θ. One would then quite naturally think in terms of probabilities over a set of population space $\Theta_1, \Theta_2, \ldots$ $\Theta_k, \ldots$, and this is possible, giving rise, as Good suggests, to probability-type chains. However, not only is the possibility of verifying that $\Theta = \Theta_k$ by any observational means, out of the question, but we also lose the advantage of the parameterization of our population spaces. The number of possible probability distributions over a set of propositions H is colossal, far transcending the ability to parametrize the whole set with a finite number of parameters. We would therefore be restricted to considering a finite set $\Theta_1, \ldots \Theta_k$ such as: $\Theta_1 \equiv$ 'set of Normal distributions', $\Theta_2 \equiv$ 'set of -ve exponential'.

Whether the population probability measures, or the populations themselves, are determined subjectively or objectively, they form the basis of the theory of 'sufficient statistics' essential for any computable theory of sufficiently long sequential processes, in which the precise nature of the stochastic properties of the propositional set is not known.[1] As with the other deductive elements brought into choice situations, they rest very much on the 'reasonableness' of the premises on which they are based. However, given this, we can considerably enlarge our powers of analysis.

[1] H. Raiffa (57).

So far we have talked in terms of probabilities, quite naturally, for they are the most developed form of uncertainty measure so far. Similar remarks apply, quite obviously, to any other measure such as Shackle's degree of surprise. For example, in the batch sampling problem, Shackle might have some surprise function, $\rho(q)$, on the batch qualities, and some surprise function on the observations from batch sampling once q was known. It is quite possible that we could have mixed inferences. We might, for example, feel surprise for the value of q, but think in terms of probabilities for the sample results.

It is interesting to note how the various theories of uncertainty are connected with the three forms of inference. The following table gives some idea of the connections. Note that the direct inference column does not apply to subjective theories since no background population knowledge is specified, whereas θ, and its interpretation, is given for objective theories. Also, in some theories, such as that of Savage, the uncertainty measures can refer to observations or population parameters, subject to or not subject to verification.

	Direct	Inverse	Predictive
Von Mises			√
Carnap	√	√	√
Classical	√	√	√
Savage		√	√
Marschak		√	
Raiffa		√	√
Shackle			√
Good	√	√	

2.4 Interdependence of Choice, Value, Uncertainty, and Knowledge

The past three sections have been concerned with theories of choice, value and uncertainty. Axiomatics is seen to play an important part in the validity of any measures derived and also in their use. Decision can be introduced into choice with the aid of these theories. Ultimately it is seen that the verification of any of these theories lies in the possibility of testing axioms of choice. Everything ultimately depends on choice behaviour in some way. So called 'objective' measures have to satisfy certain axioms of choice, before they can be used validly in any problem analysis. Although we have, in accordance with the manner in which they are covered in the literature, made an attempt to distinguish theories of value and theories of uncertainty, they are all effectively covered under the one heading, viz. theories of choice.

Theories of choice state, effectively, that under conditions c, choice will be in accordance with $\phi(c)$ where ϕ is some function on $\{c\}$. ϕ is not necessarily cognitive and until some such element ϕ has been identified[1], we are operating in an area of pure choice. c will contain background knowledge, as well as problem identification knowledge. In some theories (e.g. von Neumann's and Marschak's) c contains 'objective' probabilities. In others no knowledge content in c is specified outside the problem definition itself. From such theories, subjective uncertainties are derivable, which have their uses in two ways at least,[2] as has been pointed out. None of these theories actually delve into the actual content of c. Thus neither the compatibility of propositions in c with themselves nor with known scientific results is considered. Knowledge, as Churchman says, is only properly derivable by observing choice and hence once again our theories depend heavily on choice.[3]

It is implicit in a pragmatic decision theory that there will be circumstances in which the introduction of an element of decision would improve the selection process, and hence certain selection processes will be changed. They may simply be mechanized, and result in the same selection as the person himself would make, or they may be changed in that a different selection will be made. It is therefore important that we shall be clear as to the possible changes which might be made and in what sense they may be justified. This

---

[1] The decision-maker may not be aware of ϕ, although an observer may determine an apparent ϕ. Note that equation (1) of Chapter (i) refers to knowledge, K, and operator θ. c contains Q and that part of K which the observer knows. ϕ is the apparent decision operator on c resulting in choice $q \, \varepsilon \, Q$.

[2] Communication and use in Baysian probability analysis.

[3] We can never be sure we know all that a person knows by asking questions. Even those things he says he knows may not stand up to investigation in choice situations. This was discussed in Section (2.3.5).

we shall discuss in Chapters (4) and (6), where we shall discuss a little more fully the ways of introducing decision into choice and just what it will imply for the person concerned. Ultimately all of our uses of decision theory will depend on certain basic characterizing theories of choice. We might illustrate this simply as follows:

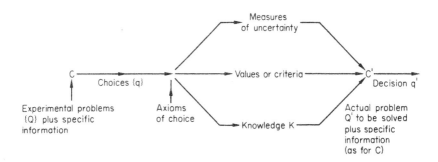

Before we can profitably enter into this discussion it will be essential to discuss a little more fully the idea of decidability and the different degrees of decidability.

Summary

The starting point for any of our theories is the observation and characterization of choice behaviour in specific situations. We develop this in the context of value and uncertainty.

2.1 Choice and Order Relations

The formal development is through the concept of 'order relations'. These order relations arise in both theories of value and uncertainty, essentially on the basis of observing choice behaviour. A major difficulty arises in Decision Theory, in the use of these, through the attempt to infer preferences by observing choice. The general properties of relations are outlined and specific consideration is given to the 'transitivity' and 'comparability' properties because of their essential roles in Decision Theory. Intransitivities occur for several reasons and seriously restrict the development of a useful Decision Theory; incomparabilities occur for several reasons and result in 'partial decision-making', and 'Pareto' optimality. Part of the task of Decision Theory is to make alternatives comparable.

2.2 Value

Value Theory is one of the major components of Decision Theory. The general idea of a real value function over a set of elements is discussed. Complications arise when the idea of probabilistic choice is introduced in order to cater for intransitivities. Probabilistic choice may be operated away by proper evaluation of alternatives, however. Consideration is given to some axioms of probabilistic choice, particularly the restriction to dichotomous problems. The existence of an ordering of alternatives does not guarantee the existence of a real value function, and we may have lexicographic values; these are considered from the point of view of actuality and validity. Constraints on problems give rise to invalid lexicographic values. In a countable bounded situation we can have equivalence between real and lexicographic values, with some computational advantage in the latter.

The mere construction of a real value function is of no value. The usefulness of value theory lies in the ability to determine the value of any element from the knowledge of the values of a few elements; we deal now with value structures with this in mind. We first of all consider Milnor's theories in the context of a general (rather than a dichotomous) problem Q and establish the difficulties which could arise if such behaviour as the Savage Minimax-Regret criterion is permitted; this prevents the pairwise comparison of alternatives and is very restrictive in the development of Decision Theory; it permits no real valuation of alternatives; the idea of a 'better decision' is not permissible.

The remaining discussion is limited to real value situations and special value structures are discussed, with particular reference to their computational benefits. The main concern is with the manner in which elements in χ combine and with the corresponding manner in which their values combine. Consideration is given to some of the axioms of choice involved.

The emphasis is on value structures and computability and two misconceived views on value theory are discussed to drive home the main purpose of value theory.

The problem of axiom verification on the basis of 'ought' or 'description' is discussed briefly.

Perhaps the most useful application of value structures is in the area of sequential problems. We deal here in terms of values of alternatives and their relationship to the values of their consequences. All processes are essentially sequential and nonterminal and this adds to the importance of this section in realistic decision situations. Comparison of the Normal and Extensive forms for sequential processes is made and the relevance of value theory to the Extensive form is demonstrated. The Normal form is impracticable in most realistic situations. The development of Value Theory for the Normal form requires rather more stringent conditions than the mere homomorphism by which Value Theory was introduced; the relationship between problems and subproblems has to be considered (or equivalently, the relationship between events and consequences). We find that, for example, von Neumann's theory is very suited to specific types of sequential situation, but that, in general, we do not even have equivalence of the Normal and Extensive forms. This constitutes a major difficulty in the development of Decision Theory.

Finally, to illustrate the essential theme of the 'event-consequence-value' triad, it is indicated how the requirements of value sufficiency for the sequential situation can be used to further develop Shackle's theory.

2.3 Uncertainty

We now discuss fully the uncertainty aspect of decision-making with the ultimate conclusion that it is essentially a subjective concept whose ultimate meaning lies in its place in describing choice behaviour.

Measures of uncertainty apparently separate into 'objective', 'subjective' and 'subjectively derived objective' measures. The objective measures are standard procedures operating on data presented formally in a standard way; the subjective measures are measures obtained by observing how a person chooses in given situations and are measures which describe actual choice behaviour; there are then those theories which stipulate that probability has an objective meaning but is subjectively derived. A variety of the various theories is explored. It is seen that the essential purposes that objective measures serve is that (a) they can operate on a mass of data, and produce information which is meaningful but which is not reliably produced by a human, (b) they provide routine methods of combining the various uncertainty aspects of a complex problem to form synthesized uncertainty measures. However, in the ultimate analysis, such measures are valid for decision-making only if they actually enable choice to be predicted in certain standard situations. If we now accept such objective probabilities we need to be able to interpret them and this may be done by identifying standard events with the same objective probabilities, so that '"the probability of E" $= \cdot 5$' means that the person would choose in the same way if E were replaced by 'a coin will fall heads'. This sort of interpretation is again subjectively oriented and avoids many pitfalls covered in the theories initially described.

A central feature in the development of any theory of uncertainty is the degree of specification of the knowledge content K on the basis of which the measures are to be derived. Some theories (e.g. Savage) specify the contingencies, but not whether the subject has particular knowledge influencing their occurrence. We then have the dilemma that, if communication between an adviser and a decision-maker is the prime purpose of the uncertainty measure, we do not know whether, in any case, the adviser's measures are good or bad; different people, because of their different experience, and hence knowledge, will have different measures. This problem is quite open. The knowledge problem is quite difficult, e.g. the manner in which even objective probabilities are derived depends on subjective assumptions about influencing factors; even if each person had the same knowledge, his state of mind at the time of the study will provide an important bias, as Shackle points out; even the celebrated Principle of Insufficient Reason, resting as it does on concepts of 'symmetry of knowledge' breaks down under examination, and the fact that some probabilities appear to stem from this principle is seen to be incorrect since these probabilities can actually be deduced from other reasoning.

Having explained the theory of uncertainty we now consider the problem of 'determinism versus uncertainty' with respect to (a) its achievability and (b) its desirability. Trends in science, in general, are in the direction of uncertainty, and such principles as Heisenberg's Principle of Uncertainty are seen to be relevant to human behaviour. The search for certainty may be neither economically feasible, nor even desirable as a human prerequisite. Even the possibility of certainty achievement, given enough time and resources, is questionable because of the openness of the system being studied as distinct from the relatively closed nature of some of the physical sciences studied so far. We consider the so-called 'physical versus human science' comparison later on. Finally we discuss the problem of 'necessary versus unnecessary ignorance' in retrospect of a decision.

So far we have been concerned with the uncertainty of an individual proposition, whereas there are measures of uncertainty for a collection of propositions as a whole. Since these do arise in the discussion of subject matter close to decision-making, we examine them and conclude that the concept of 'entropy', in particular, has no direct bearing on Decision Theory, and is solely concerned with information processing capacities; indirectly, however, it does influence general considerations of computational capacity. The other collective measure of uncertainty is the 'variance' of a population and this is of essential importance to Decision Theory and plays a part in describing the risk characteristics of individuals.

Finally, we consider the three main forms of inference which arise in Statistical Decision Theory, viz. Direct, Inverse and Predictive Inference. These are seen to hinge on a combination of subjective and objective probability measures. Their essential purpose is the combination of initial subjective measures with the collection of ongoing formal observations to produce, formally, new probability estimates and thereby combine subjective experience with the reliability of formal analysis.

2.4 Interdependence of Choice, Value, Uncertainty and Knowledge

It is seen that all our measures stem from some theory of choice. The axioms of choice play an essential role in allowing complex problems to be decided on the basis of choice behaviour in simple standard situations. The behaviour need not be cognitive (to the subject), but is still a characteristic of use in decision-making. Such characteristic behaviours depend very much on the form of the standard conditions, and the problem of knowledge is of paramount importance in such theories, yet to be explored more fully. Finally the very use of any of our theories will change some choice behaviour—if not then there would be no point in our study. It remains to consider the ways in which the changes can occur and be justified.

CHAPTER 3

Decidability

Chapter (2) discusses value and uncertainty and their dependence on choice behaviour. The choice behaviour was in no way necessarily connected with decision. In this chapter, we shall discuss the content of mathematical (or logical) decision, and several problems arising in the introduction of decidability into choice. It is necessary to note that we can permit several forms of decision, depending on whether it is complete, partial or probabilistic. Although we talk in terms of probabilities, any other valid truth value measure would do.

3.1 Complete, Probabilistic, and Partial Decidability

Decidability, as used in the true mathematical sense, is concerned with the 'proof' of a proposition, or its negation, based on a set of axioms and the usual two-valued logic. Thus, in geometry, we can 'prove' or 'disprove' propositions about intersections of lines, given the properties (axioms) of the manifolds with which we deal; in the field of number theory we can 'prove' certain propositions about sets of numbers, given the arithmetical axioms. The important point here is that of 'proof' and its identity with 'decision'. Dunlop (27) stresses this identity. In such usage, of course, our problems contain only two alternatives, viz: truth or falsity of the theorems proposes. The axioms may, however, not be sufficient to determine truth or falsity. In this case we have three alternatives, viz: true, false, undecided. However 'undecided' just reproduces Q and need not be considered separately. Thus, if, in convergence analysis, we take as an axiom that, for the series $u_1 + u_2 + u_3 \ldots u_n$, $|u_n| \leqslant k\,|u_{n-1}|$, $0 \leqslant k < 1$, then the problem is $Q \equiv \begin{bmatrix} T \\ F \end{bmatrix}$ and the selected alternative is T, i.e., the series will converge). If $k > 1$, we are undecided.

Consistent with this usage we say that we have complete decidability for a class of problems if decision criteria exist which determine, for each problem, one, and only one, act to be selected.[1] It

---

[1] We have already stated that the process of problem solving is a decision process itself.

must be remembered that such criteria may result in several stages of analysis and selection prior to the final primary selection. It must also be remembered that the choice must be from the set initially identified. Thus, if we start off with a problem Q containing two alternatives, and others are generated in the process, then Q would have to be re-identified with the two alternatives plus the residual set, Q', to be identified by the decision criteria.[1] Thus, if a manager puts forward two alternative proposals to his director, the selection is to be decidable, and if neither proposal suits the director, then the decision criteria should determine under what conditions, and how, to determine further proposals. This is asking quite a lot, but, within the context of our mathematical decision, if the selection is to be completely decidable, this must be adhered to.

In Chapter (2) we discussed measures of uncertainty as truth values on certain propositions. There is no methodological difference between accepting a proposition as true and selecting a certain act to be taken,[2] the former being a commitment to act, in future, as if the proposition were true. We can thus consider an extension of the certainty form of mathematical decision to the probabilistic form, and this ties up quite naturally with probabilistic choice, although we shall require a cognitive probability mechanism to be identifiable. The classical probability calculus then permits statements of the kind '$A \to h$' to be extended to '$A \to$ "$T(h) = p$"' where A contains probability propositions as part of its premises.[3]

This approach introduces a problem of interpretation, of probabilistic decision, in view of our previous remarks about the equivalence of assigning truth values to propositions and selecting an act. If $Q = (a_1, a_2, \ldots a_m)$ is a problem, and if we introduce probabilistic decidability to derive $p(a_i) = p_i$, we can either interpret this to mean 'the probability, that a_i is the correct act to take, is p_i' or 'act a_i will be taken with probability p_i'. There is a significant difference between the two usages, which we will illustrate below. Both have

---

[1] In other words Q should always be so defined as to allow for identification of other alternatives later on. Thus $Q \not\equiv \begin{bmatrix} a \\ b \end{bmatrix}$ but $Q \equiv \begin{bmatrix} a \\ b \\ Q' \end{bmatrix}$ where Q' comes from the decision process.

[2] If we reduce the problem Q to a problem Q' such that the person is really indifferent between each pair of alternatives in Q', he still has to make a choice and we would need an operation to accomplish this.

[3] A is a set of propositions, e.g. axioms.

valid uses, in appropriately defined conditions, and both are relevant to decidable choice. The first usage would require a clear idea of what is 'right'. The second usage would mean that von Neumann's expected value theory would have to be replaced by Luce's theory of probabilistic choice (the theory of games does not overcome this because two different sorts of criteria are brought to bear (see Chapter (7), Section (4)). In the deterministic case they are identical.

It is quite common to hear, from both businessmen and academics (Brown (14)), statements of the kind 'we wish to increase our chances of making the right decision'. It all depends on what is meant by the 'right decision'. Consider the following, single stage, example. If we say that an act is the 'right' one if it results in the maximum return, then act b will be chosen, given the premise implicit in the definition of 'right', provided the 'probability of being right' is a sufficient indicant for choice. The matrix entry is the probability that act i will result in outcome j.

		outcomes			
		1	2	3	4
acts	a	·1	·2	·4	·3
	b	·6	0	0	·4

This is consistent with the first form of probabilistic decidability. It must be realized, however, that it is possible for neither a nor b to result in the maximum return, and that $\sum_{a\varepsilon Q} p(a) \neq 1$ in general.

If we are concerned with choosing which of a given set of propositions to take as true, then an act is 'right' if the proposition corresponding to it is true. For a two proposition problem, $(h, \sim h)$ we have the following format.

		States	
		h	$\sim h$
Q acts	h	1	0
	$\sim h$	0	1

At least one, and only one, act is right and hence $\sum_{a\varepsilon Q} p(a) = 1$.

The consequences are represented by '1' or '0', but in practice we need to evaluate exactly what these are.[1] Thus '0' may not be seriously undesirable, and may be little different, in effect, from '1'.

The use of 'right decision (act)' gives rise to even more severe difficulties for multi-stage processes. Consider an n-stage process which moves from one state to another in a finite set, and suppose that a sequence of moves is 'right' if the system ends up in a target state 1 after n stages. Let the acts be numbers $1, 2, \ldots k$ at each stage, and that, at each stage, a 'state', i, is 'right' if it maximizes the chances, under an optimal policy, of getting into state 1 at the end. Let $p_m(i)$ be the probability of getting from state i to state 1 in m stages using an optimal policy. Then, if $p_{ij}{}^k$ is the probability of moving from i to j when action k is taken, we obtain the following equation.

$$p_m(i) = \max_k[\Sigma_j p_{ij}{}^k p_{m-1}(j)]. \tag{2}$$

Now it seems reasonable to require that, if $\max_j[p_{m-1}(j)] = p_{m-1}(j^*_{m-1})$, then j^*_{m-1} is the 'right' state to be in at stage $m-1$. However, it is then not true that acts should be taken to maximize the transition probabilities into 'right' states at each stage since for the optimizing k^* given by equation (2), k^* does not satisfy, in general,

$$\max_k[p^k_{ij^*_{m-1}}] = p^{k^*}_{ij^*_{m-1}} \tag{3}$$

The main point is that probabilistic decision is allowable provided we allow probability type deductions, and a simple extension of the usual dichotomous truth values. However, there are two interpretations of '$p(a_i) = p_i$' and an extension of the concept of the 'probability of a proposition being true' to the concept of the 'probability of an action being right' has certain dangers. In the deterministic case there is no conflict.

It is a common occurrence in a selection process to have limited decidability in that, at some stage, a problem exists, the resolution of which is by means of pure choice.[2] This may arise because a particular stage is subject wholly to choice. Thus, the set of alternative actions to be considered may be purely subject to choice, or the

---

[1] There could be four different consequences of course. If a consultant has to decide whether a person has Thyroid Cancer or Hashimoto's Disease (excluding other possibilities), and the treatment and prognosis are predetermined as a function of his decision, then the consequences of treatment for Thyroid Cancer when he has Hashimoto's Disease will be different to the consequences if the reverse situation pertains.

[2] It is worth repeating that by 'choice' we mean the act of selecting without any formal expression of the reason for doing so.

information to be obtained for a specific problem may be purely subject to choice. It may arise by reducing the initial problem to a smaller one for which no further discrimination is possible: thus, given a set of candidates for a post, we may be able to decide that certain candidates can be ignored because they do not fulfil certain prerequisites. Alternatively at least one other candidate may be better in all relevant respects. Beyond this it may not be possible, within present capabilities, to decide the ultimate selection. Similar cases exist where alternatives are 'dominated' in almost all problem areas. Thus, if a labour force gets beyond a certain size, we get no benefits from any increase and can hence remove all problem alternatives with a labour force beyond this size.

This gives rise to partial decidability. Partial decidability allows considerable reductions in the size of problems. Complete and probabilistic decidability removes the problem altogether.

Although a problem Q can be reduced to a problem $Q' \subset Q$, in which further discrimination is not, at present, possible, it does not mean that the person is indifferent to all $q \, \varepsilon \, Q'$. It simply means that we have not established any premises[1] which would enable us to resolve Q and hence, at this point, we have to leave this final selection to him. It is possible that the person would be indifferent to all $q \, \varepsilon \, Q'$ even if he could discriminate fully in Q'. This occurs in mathematical optimization problems in which several solutions give the same optimal value. Such situations can easily be made decidable by introducing any arbitrary rule which will guarantee a single selection. This might be by means of probabilistic decision, with specified probabilities. We need to be sure that indifference is real, and hence that the decision criteria are as complete as they possibly could be.

It is not to be assumed that complete decidability is to be sought after. As with the quest for certainty, it is a costly process to achieve this even if it is possible. Thus the effect of quality on lost sales may conceivably be determined given enough time and money, but it is likely that an assessment of this will be left to the 'expert'. We may end up with

$$Q \equiv \begin{bmatrix} a_1 \\ a_2 \end{bmatrix} \text{ where } a_1 \equiv [(\text{inspection cost})_1, (\text{quality})_1]$$
$$\text{and } a_2 \equiv [(\text{inspection cost})_2, (\text{quality})_2].$$

---

[1] These premises may be related to refinements of choice behaviour or to such things as more information on the consequences of some of the alternatives in Q'.

The resolution of Q may then be solely a matter of choice.[1]

The whole process of 'delegation' in an organization arises because of the benefits from making certain problem areas only partially decidable at each stage of its resolution. Thus, in an insurance business, critical claims levels are set which 'decide' which person will handle the claim, but not what the outcome will be. A manager may have rules determining which problems he will handle and which problems his subordinate will handle, but not how they will be handled.

Having differentiated between the different types of decidability let us now consider the means of achieving decidability. The central role is played by axiomatics and we shall discuss these.

3.2 Decision and Axioms of Choice[2]

Mathematical decision is founded on the axiomatic approach in which the truth or falsity of propositions may be deducible from a basic set of postulates (or axioms) themselves assumed to be true. However, although the only prerequisite in mathematics is that there be internal consistency among the axioms (i.e. no deductions can be made from a subset which conflict with another subset), in theories of choice we are concerned with the validity of such axioms as well, even if they are internally consistent. The search for an internally consistent set of basic valid axioms of choice behaviour is the foundation stone of decision theory.

The axiomatic treatment of theories of choice is intended to bring out general characteristics of certain sorts of situation. The development is through the ideas of actions, consequences and events, and is not oriented towards choice behaviour in specific real life situations, although these can be given interpretation within the general theory or even deduced from the general theory. Milnor, for example, tries to cater for a general class of problems in which we have 'symmetry' of information. Savage's viewpoint is that all problems

[1] We may be able, by observing choice behaviour, to construct a criterion function, v (inspection cost, quality) and thereby make the problem decidable. If it were found that the inspection cost was a function, $f(q)$ of quality, q, then the solution would be decided by finding q to maximize $v(f(q), q)$.

[2] An axiom of choice is a general proposition relating to a person's choice behaviour in specified problematic situations. It is of the form 'given c (containing Q) then he will choose $\phi(Q)$, where ϕ is an operator'. ϕ may not be cognitive to the decision-maker. A criterion function may be taken as an axiom of choice, of a high order.

fit within the subjective probability form. Whichever theory is correct, the precise criteria, values, and uncertainties factors can be determined for specific situations. The axioms themselves may be accepted as being ethically correct, or they can be tested as being actually correct for a specific individual.

As with mathematics where the search is for the minimal set of postulates from which a large number of deductions can be made, and, as in science, in which the search is again for a minimal set of laws of nature from which considerable behaviour can be predicted, so it is in practical action selection processes that the derivation of an adequate set of axioms of choice will make a very wide class of problems decidable in principle and computable in practice.

The axioms mentioned in Chapter (2) are of different types being either of the topological, mixture (or composition) or order and group operation types.

Topological axioms such as Chipman's 'well ordered' axiom requiring the set χ to be indexable by an ordinal number, or Debreu's 'separability' axiom requiring the existence of a countable subset whose closure is χ, are axioms which essentially refer to the ability to identify elements with numbers of one kind or another. Such axioms are relevant to measurability in general. However, because of the inherently discrete nature of all our measures, these axioms have only a mathematical interest, since the existence of our measures follows directly from this discrete nature.[1]

Mixture axioms such as von Neumann's axiom that '$a,b \, \varepsilon \, M \rightarrow apb \, \varepsilon \, M$', or Ackoff's requirement that '$x,y \, \varepsilon \, \chi \rightarrow x + y \, \varepsilon \, \chi$, essentially give us some idea of structural composition properties of our set χ.

Order axioms say something about the properties of choice in given problematic situations. The transitivity axiom requiring that '$(a > b, b > c) \rightarrow (a > c)$' and most of von Neumann's axioms are essentially order axioms.

The group operations axioms, which play an extremely large part in decidability, require group operations over χ to be related to the group operation on the real line or lexicographic space. Thus we

---

[1] The fact that the interest is purely mathematical does not mean it has no value. Classical calculus approaches to solving problems are powerful since they inherently exclude a large area of search from consideration. However, they do not apply unless we have continuity and differentiability properties. Debreu's lexicographic example in Chapter (2) shows that it is conceivable that no continuous value function exists. Also the fact that irrational numbers do not exist in practice does not preclude us from using formulae for optimum solutions (see p. 163).

have Ackoff's additivity axiom that '$v(a+b) = v(a)+v(b)$' and von Neumann's postulate that '$v(apb) = pv(a)+(1-p)v(b)$'.

The central axiom types are the order axioms and group axioms which are directly concerned with choice and verifiable only by observing choice. They relate to a class of problems, (Q), over a set of alternatives χ, in a given informational context.[1] It does not follow that axioms concerning choice in sets of dichotomous problems are sufficient to determine choice in multi-alternative problems (see notes on Savage's minimax regret criterion). We may, however, wish to assume, on 'reasonable' grounds, that this is so, and hence have a value function consistent with choice in all Q over χ, and then the axioms need only refer to dichotomous choice situations.

A brief review of the axiomatic approaches in Chapter (2) will show that some axioms are very general in nature in that no identification of the elements $x \; \varepsilon \; \chi$ is made. Some give partial identification only in as much as x is a vector of 'commodities' (which is quite general). Real (or apparent) decision rules in investment problems, inventory control problems and personnel problems are expressed in terms of a more specific language describing the specific type of problem considered. Within our decidability framework either language is acceptable providing it allows us to decide. The advantage of the general language (such as that of von Neumann mentioned above) is that it may be applied to several classes of problem. Thus although choice behaviour in the areas of inventory and investment is expressed in a different language von Neumann's axioms may apply to both problem areas. In the former case, an alternative may be specified in terms of $x \equiv$ (cost, probability of run-out) $\varepsilon \; M$, whereas, in the latter case, an alternative may be specified in terms of $x \equiv$ (probability distribution of cash) $\varepsilon \; M'$. It may be that von Neumann's axioms apply to M and M', in which case we may construct value functions using this theory. Alternatively, we may simply observe choice in M (and M') and construct valve functions directly on this basis—bearing in mind that it would be of no real use unless some value structure were evident.

We can run into difficulties when considering classes of related specific problems. For specific classes of problem we can observe choice and hence the validity or falsity of any axiom we may wish

[1] Axioms can be expressed in general or specific languages, although when reference is made to the axiomatic theory of choice it is usually to the general theory to which reference is made.

to test. We may find sets of axioms which fit behaviour for various classes of problem. If we are simply trying to describe choice then there are no difficulties. However, it is a premise in our research that certain choice behaviours are to be taken are inviolable and others are to be changed. Which behaviours are to be accepted as basic to our analyses, and which are to be changed?

Consider the choice behaviour when a person is faced with a set of alternative investments in which direct cash flows are given for each year. We may find that he always chooses those in which the net present worth, using a specific discount factor, is largest. On the other hand if he knows the probability distribution of all resulting cash flows he might choose differently, following, perhaps, von Neumann's theory. There are, therefore, two distinct sets of axioms of choice depending on the form of the problems presented. The first one considers the investment as a time series of direct cash flows and operates on this time series. The second one considers the future decisions on the use of these cash flows, and the consequential effect on total cash flow, and operates on the probability distribution of total cash flow using a von Neumann-derived value function for total cash at the end.

The answer to this must lie in realizing that some special[1] axioms may themselves be decidable providing the problem sets are connected.

General axioms of choice can be used to determine special decision rules. Thus an investment decision rule may be used, given the nature of the behaviour of the environment within which the problems are generated, to generate propositions about the financial consequences of the rule. The acceptance or rejection of the rule then depends on the acceptance or rejection of the characteristics of the consequences of the rule as determined by the general axioms relating to the consequences.

In Chapter (2) we mentioned the problem of consequences of an act, and Savage's query as to whether indecomposable consequences exist. The basic axioms from which all other specific axioms (and here I am referring to propositions relating to criteria of any kind used in problematic situations), may be decided should be in terms

---

[1] Thus the use of discounted cash flows is specifically related to investment-type problems presented in the cash flow form, whereas if everything were reducible to probability distributions of final cash we might have a general criterion based on value functions for probability distributions of cash.

of indivisible consequences. In practice this may create difficulties and it may be uneconomic to try to do this. This is a serious difficulty, limiting our deductive content, for, to some degree or other, all problems, arising at any point in time are connected. The problem environment is connected in that the solution to one problem may not only generate such problems but may also influence the choice in future problems. Thus the consequences of choosing a particular investment not only depend on the solution to future problems posed by the use of capital needed now, but also the physical nature of the investment generates design and control problems.

One theoretical way out of this would be to try to specify a complex hierarchical policy, π, which determines precisely what to do whenever any problem arises and then to determine, for the given problem environment, propositions in terms of indivisible consequences such as money (if we wish to assume this is indivisible). We then use general axioms of choice concerning these consequences to determine an appropriate π. This is the 'Normal' form mentioned in Chapter (2) and is, within the economic limits placed upon selection processes, completely out of the question.

The alternative approach is to use the 'Extensive' form in which only the initial problem is considered, in which the 'consequences' are evaluated 'somehow or other' and in which choice based on this consequence representation. The consequences will be decomposable in principle, and it may be that subjective evaluation does account for its decomposability and their dependence on future events and problems. Axiomatic approaches, such as that of Milnor, deal with representations of this kind in which the consequences of an act have 'values' which represent all future behaviour relative to indivisible consequences. However, it was pointed out in Chapter (2) that such values may not be sufficient for the problem on hand. If care is taken to see that such values are valid then the Extensive form will be the most realistic one to use. Ackoff (2) uses this approach when the initial problem is to choose between certain company investment opportunities, in which the consequences are, among others, stability of employment, customer good will, personnel satisfaction. These are theoretically decomposable, but no usable procedure exists, at the moment, for evaluating the consequences of these in terms of money. A subjective valuation[1] is introduced at this

---

[1] This is referred to as a 'value judgement'.

stage to overcome this difficulty. The axioms of choice are those relevant to the consequences presented in this form.

We thus see the significance of the axiomatic approach to general choice behaviour as a means of determining decision rules in the widest possible class of problems. It adds considerable decidability to specific choice situations.

Let us now consider the other main component in adding decidability to choice situations, viz: measure.

3.3 Measure and Decision

3.3.1 Meaning and Suitability of Measures

Chapter (2) has been solely concerned with the 'measuring' of values and of uncertainties. These measures are simply homomorphisms of the set of alternatives, in the first case, and of propositions, in the second case, into the real (or lexicographic) space of numbers. However, this in itself is of no use. For such numbers to be of any use they must carry with them certain properties. We must be able to use these numbers as premises in some problematic situation to deduce certain results.

We discussed how value and uncertainty structures could be used to determine choice or composite uncertainty measures in complex problematic situations. Thus if $x = y + z$, then we might have $v(x) = v(y) + v(z)$ and hence be able to say something about whether x would be chosen in a specific problem.

Objective probability measures, are, as we have seen in Chapter (2), not independent of choice. '$p(E) = \cdot 2$' must be interpreted by finding a simple event F whose probability (in a specific sense) is $\cdot 2$, and then saying that E and F are equivalent in all problematic situations. This is a specific operational interpretation of a given probability measure.

One school of thought, as pointed out in Chapter (1), is that a concept's meaning (and hence its uses) lies in its operational derivation, and thus the proposition '$m(e) = \alpha$' has no meaning beyond that implied by its operational derivation. Another school (see Ackoff (1)) believes that the meaning of a measure lies in the deductions[1] which can be made using it. Thus if we say that the weight of

---

[1] Allowing probability statements.

x is 1,000 lb. we would deduce that we could not carry x. Both points of view are correct, and it all depends on whether the properties of the measure can be determined from knowledge of its derivation or whether predictions based on the assumption of such properties can be observed to be true. There exist methods of measuring a person's problem solving ability, in which the operational derivation is specified by presenting a person with a finite, specific, set of problems, and applying some rule to the results to get an I.Q. measure. The use of such measures lies in the ability to predict problem solving performance in general classes of problems. To be correct we can only say 'the I.Q. level was determined by observing behaviour in a class of problems (Q)'. To be able to infer that this measure had predictive powers in a general class of problems we would have to test it experimentally.

The danger of using measures for problematic situations for which they are not suited was illustrated in Chapter (2), both in respect to the uses of values in sequential problems, and in respect to the sufficiency of probability measures for choice (where two events with the same logical probability, but belonging to different sets of events, may conceivably not be interchangeable in choice situations).

The measurability, and validity of such, of an element, will depend on the nature of the property we are trying to represent. Some measures we accept as basic. Thus we can measure the 'cost' of some items, we can measure the 'man-hours' required for certain jobs, we can measure the quantity of material required for certain jobs. There are, for these measures, standard operational methods for doing it. Such measures may convey the full content of the information relevant to the problem on hand. In some cases the full content of relevant information may require several measures. Thus cost, delivery time and quantity of material may all be required for a particular problem. We can deduce certain consequences of an action when such measures are given, or we may be able to deduce what action to take, such as to borrow so much money or to order so much material.

Other measures are not so readily available and considerable thought has to be given to them. For any problematic situation there are many measurements one can make, since there are a variety of factors one may identify in any situation. Take for example our standard problem of evaluating a person who is applying for a job. There are many forms of information about the person which may be relevant to his suitability for the post. Thus his qualifications, age, experience and intellectual ability may be relevant. Not only is there a severe problem of identifying the determinant factors, but there is real difficulty in measuring some of these factors themselves in such a way that the measure contains the full content of information from which they are derived relevant to the problem on hand. Measuring 'ability' is no different to measuring uncertainty. It is a way of condensing the body of information describing a person to a comprehendable form sufficient for conveying its content relevant to the problem on hand.

Just as value and uncertainty are complementary measures in choice situations, so do measures such as intellectual ability have complementary factors. Intellect alone may be insufficient information on which to base the selection. Our objective is to introduce an element of decision and, ultimately, the appropriateness of any measure will depend on what we can deduce from it. Thus eventually one provides a set of measures, $[m_1, m_2, \ldots m_k]$, and the decision content lies in what one can deduce from the combination of this set of numbers. To be able to deduce anything in any experimental set up they cannot be treated separately and any other factors defining experimental conditions should be specified. To be strictly correct, one ought to model the experiment in exact conformity with the problem environment and economic environment in which the person will be operating.

Separate experiments will be necessary to determine $[m_i]$ in the first place. The procedure will be to define the operations for determining the m_i's, to measure the m_i's and then to observe behaviour in the specified environmental conditions.

Thus we might define personnel satisfaction (m_1) on the basis of the score on a given questionnaire. We might define customer satisfaction (m_2) on the basis of a similar score. We then need to observe the effect on company performance of various levels of m_1 and m_2. Such approaches, which begin with definitions of the measures and then test their predictive power, run the risk of giving negative results, since the measures may not be the appropriate ones.

At the same time, the axioms can be intuitively chosen, giving rise to less chance of negative results in the end. This is what has been done in Chapter (2) in relation to value and uncertainty. Such approaches can be used to determine measures on investment opportunities, under given conditions, which allow appropriate investment problems to be made decidable. The single measure (of 'present worth') will be sufficient to determine choice under certain conditions (White (72)).

It has been stressed that we are only interested in measure in as much as it allows deductions of one degree or another, in the context of problematic situations, and that, although many measures of an element are possible, only certain ones are meaningful for the problem on hand. Common measures which raise serious problems in this context are those of 'profit' and 'capital value'. Many differences of opinion exist concerning the 'appropriateness' of any operational determination (Churchman (20)). No satisfactory resolution is likely to arise unless they are examined against the background of the problematic situation for which they are designed, whether this be for the Managing Director of the company or for shareholders, present and prospective, or possible purchasers, or even competitors. There is absolutely no point in measuring 'profit' unless it influences the resolution of certain problems. Even as an 'index' of performance it serves the purpose of influencing the solution of the class of problems 'whether or not to make changes in the business operations'. It does not, of course, predetermine what actions to take, and is quite likely to be subjectively evaluated in this respect. To be able to determine what action to take we have to deduce, fully, partially or probabilistically, the consequences of taking any action contemplated, given the 'profit' measure. This may require inductive arguments of the kind 'if it is allowed to continue as it is, then . . .'. It may make no allowance for the possibility of choosing differently next year, or for circumstances to change, and is hence a particularly weak form of measuring. It is, nonetheless, existent.

There are different ways of measuring profit and of measuring capital, and it would be out of place to discuss these here. It may be that there are different purposes in mind when doing so, depending on whether it is purely for accounting purposes, whether for taxation purposes, or whether they are regarded as indicants for operational problem resolution. If so then they ought to be labelled as such. Such measures will then be relative to the class of problems for which they are designed. Thus 'profit = income − expenses' may enable the tax collector to solve the problem of how much tax to charge to the company. Such a measure would be virtually useless for the Managing Director who wishes to assess company performance in order to help him select any appropriate remedial action, since capital factors, influencing future company performance, may have been ignored.

Cost measures also face the same difficulties, particularly when it comes to the inclusion or exclusion of overheads. It is incorrect, for example, to include the full weight of overheads if costs are to be used to determine whether or not to sub-contract work. The answer to this depends on future decisions about the present plant. If we are going to incur the overheads anyhow, for a long time to come, then, using discounted cash flow, the cost now needed for this purpose contains very little of the overhead. If the sub-contracting will allow us to dispense with certain overheads then they have to be included in the cost.

To illustrate the pitfalls of measure in another area, consider the problem of job valuation, quite common in industry, in which a points system allocates points for each attribute of a job and the final score is used as a measure to determine salaries. The adequacy of the measure again depends on the purpose of the measure. It could be that the problem in the mind of the person concerned is that of maintaining an 'equitable' relationship between salaries and what the employee puts into his job. Thus time, effort (physical, mental), expertise . . . may be the inputs. It may be the problem of maintaining an 'equitable' relation between salaries and what the company gets out of the employee. The two may only be loosely correlated. In both cases the measures are used in resolving the class of problems 'how much shall we pay employee E?', and one can expect that the answers would be different for the two different motivating premises.

The problem of measure and its uses in problem resolution is quite a complex one. The preceeding notes by no means solve it. They rather indicate the nature of measure, why it is needed and some of the pitfalls which have to be avoided. Whether measures are subjectively obtained (operationally) in the first place and their sufficiency, as indicants of behaviour, is to be tested experimentally, or whether they can be derived from certain behavioural axioms, will depend on the context of problems we wish to resolve. In some cases we may feel one approach is feasible, and, in other cases, feel that the other approach is feasible.

Let us now turn to a discussion of the different forms of measure, in a general sense, which can play a part in a theory of decision.

3.3.2 *Forms of Measure and Ambiguity Content*

Ackoff (1) discusses 'measure' quite fully and one can refer to this for a detailed treatment. The different basic forms of measure between which he differentiates are numbering, counting, ranking and measure in the 'restricted sense'. Briefly these are defined as follows.

A. NUMBERING

Given a set of elements, χ, we simply index them with one of a set of symbols, $\mathcal{S}$. Thus we may map (red, green, blue ...) into $(1, 2, 3 ...)$ or into $(A, B, C, ...)$. The usual convention of a 'number' is not strictly adhered to, nor is it necessary. Another example might be the identification of several different pieces of complex equipment by numbers such as $X_{12}B$, $X_{34}A$. Numbering is simply a coding of elements in a set.

B. COUNTING

Whereas numbering is relative to an individual element, counting is related to a set of elements. Thus 'there are k elements in S' is a property of S. The relevant property is that sets with the same number of elements can be put in a one-one correspondence (see Birkhoff (10)). The usual relation of the form '$n > n'$' is to be interpreted as 'n, n' are numbers identifying two sets (see definition (A) of 'number') S_n, $S_{n'}$, and $S_{n'}$ can be put in a one-one correspondence with a subset of S_n'. All addition and multiplication of such numbers stems from these correspondences.

C. RANKING

Given a set of elements, S, and an order relation $>$ (see Chapter (2)) which is a chain order, then the elements can be numbered $n = 1, 2 ... m$, such that, if $s > s'$, then $n(s) > n(s')$ and vice versa (see definition (B)).

D. RESTRICTED SENSE

This is defined to be measure by the use of standards. It splits into 'interval' type measures (such as measures of length and weight) and 'ratio' type measures (such as density and frequency probabilities).

In the former 'counting' is implicit, and, in the latter, the derivation of 'rationals' from the integers is implicit.

Measures can be vectorial, but the individual components still have to be ascertained, independently of the rest, and we shall, therefore, confine ourselves to these.

We now dispense with indexing in terms of general symbols (which can, in any case be transformed into conventional numerical form). For our purposes measure is the assignment of integers, rationals, or even irrationals, to elements in a set χ.

Our essential concern is with what can be deduced from such measures in any problematic context, and our decision content will range from complete decision[1] to various forms of partial decision, depending on the degree of ambiguity in the proposition '$m(e) = \alpha$'.

Numbering is fairly trivial and yet nonetheless important. It allows a limited class of problems to be made decidable. Within the context of our definition of decision, any information permitting selection to be decided is relevant. Thus, if we want a replacement of some element we just quote its number. We do not need to specify its requisite function or properties. On the other hand, if we know its number, we can ascertain any relevant properties we wish to. Thus such a number allows individual elements to be selected from a set of elements or for properties of an element to be determined. No meaning is to be given to numbers with no associated element. Such numbers may only partly determine the element, but, even here, it serves the purpose of reducing risk if the rest is a matter of choice, or reducing search time if the 'right' item has to be found by trial and error.

Counting provides information which makes a different sort of problem decidable. If, for example, we want to know how many boxes to purchase, we will want to know how many items we are selling. If we want to know how many components to purchase, we will want to know how many products we are going to make. Such considerations illustrate very well the intrinsic property of one-one correspondence in counting. Counting measures may still leave ambiguity in choice, even if we know how many items are required this month. Thus they may not predetermine how many components to purchase now because of the dynamic setting of our problem.

---

[1] Strictly speaking we can never be sure that one number is greater than another, for one of several reasons. 'Complete' therefore is to be interpreted in accompaniment with a very high degree of probability.

Such counts may enable the admissible area of choice to be reduced however. Thus we will know that we need at least a certain number of items.

Ranking as a form of measure has been given considerable attention within the fields of economics and business in general. In order to rank a set of elements some concept of 'order' among the elements is necessary in relation to some property. Thus, given a set of applicants for a job, we may try to rank them in terms of 'ability', in terms of 'expertise', in terms of 'potential', without trying to determine restricted measures. There are advantages in the cases where only the relative order is required. Thus if we only want to say 'A is more able than B' then why measure A or B, in this respect, in any formal manner? However, such propositions suffer not only from ambiguity of meaning, but also from ambiguity of number, assuming such a number did exist. Without any final specification in terms of some number or numbers there can be no definition. All definition centres ultimately on a mapping into a set of numbers. The ability to communicate depends on this mapping. For a person to say 'A is better than B' means nothing more than that he would choose A if given the problem $Q \equiv \begin{bmatrix} A \\ B \end{bmatrix}$. It says nothing else. For a person to say '$m(A) > m(B)$' communicates something when m has a common operational meaning. However, we have to realize that ultimately propositions of the first kind form a basis for decidability, just as much as do those of the second kind. When we cannot provide objective ways of determining whether one person is more able than another then we rely on subjective assessments.

Suppose, for example, the work's engineer was presented with the proposition, relating to equipment,

     (i) A more reliable than B
    (ii) A less costly than B
   (iii) A less life than B
   (iv) A more suitable for labour than B

What help does this give him in his problem of selecting between A and B? If 'suitability' were defined (and this means some number has to be attached), if (iii) were reversed and if (i)–(iv) were sufficient for decision, then we have $A > B$ if we also assume that reliability, cost and life are unambiguously measurable. We only have to negate one of the first two requirements to reduce the problem to one purely of choice, unless lexicographic values are valid for this problem. If we had alternatives A, B, C, we may be able to 'decide', that certain ones (using the principle of dominance), may be neglected, but beyond this we may end up with several undecidable alternatives.

If all of the propositions (i)–(iv) have restricted measurability in principle, it may be possible to 'decide', without accurately measuring, providing the difference of the numbers for A, B, relative to some of these attributes offset any possible realizations in the rest. Thus suppose we have the following measures where we use r, c, l, s for (i), (ii), (iii), (iv) respectively.

$$m_r(A) = 1, \quad m_r(B) = 0 \tag{2}$$

$$\alpha \leqslant m_t(A), \quad m_t(B) \leqslant \beta, \quad \text{for } t = c, l, s \tag{3}$$

Let choice be in accordance with a valuation function as follows.

$$v(X) = a_r m_r(X) + a_c m_c(X) + a_l m_l(X) + a_s m_s(X) \tag{4}$$

Then, providing $a_r > (\beta - \alpha)(a_c + a_l + a_s)$, we always deduce the following:

$$A > B \tag{5}$$

A similar situation arises when the choice from two alternatives depends on the probability of a given event. It may happen that $A > B$ iff $p(E) > \cdot 1$, and, without trying accurately to determine $p(E)$, one might be quite sure that $p(E) > \cdot 5 > \cdot 1$. Thus we may yet be able to decide without certain formal measurements.

Thus although we cannot deduce anything unless the elements do have unambiguous numbers in principle, we can still decide some problems without having to ascertain all of them accurately.

One of the real difficulties with a ranking procedure is that it is always relative to a finite subset S, of the possible space χ of elements since one cannot rank the whole of   if it is large. This severely restricts the generality of decision theory for we can have no use for numbers derived from, and heavily dependent on, a subset S of the whole set χ. On the other hand, practical decision theory requires that we can determine the relevant measures with the consideration of such subsets. In von Neumann's theory we need only two datum elements to get the values of the rest.[1] In Shackle we have concepts

---

[1] The dependence of values on the datum subset in von Neumann's theory is irrelevant to the final results, but the dependence of ranking on the subset S can have serious effects on any theory using such derived rankings.

of 'value', v, 'surprise', ρ, 'stimulus', ϕ, and 'choice function', χ, which purport to be co-determinants of choice in any problematic situation. There is considerable argument as to whether ranking (ordinal) or restricted (cardinal) measures are valid. Gorman (33), commenting on Carter's article (17) on Shackle's theory, says:

'There is always a problem in naming classes determined by these criteria. Since, however, they are all three ordinal we can choose to label each with a higher number, higher numbers going with higher classes.'

If there were only a finite number of classes for ρ, v, ϕ, χ then it is conceivable that a ranking of the complete sets could be carried out mentally. Every alternative would then be identified with one of the χ levels, and Gorman's remarks would be true. However, if we allow a very large number of discriminated classes, this is just not practical. Therefore, we may have to resort to ranking the elements within the finite context of the problem on hand. The ranking numbers would then be dependent on the reference set S generated by this problem and would have no place in a general theory such as Shackle's theory.

The answer to this problem again depends on the degree of decidability we want in our problem solving.[1] It is quite possible, given 'reasonable' axioms, to produce a large number of discriminating values of χ. In this case ranking procedures are inadmissible if the theory is to be general, since we can surely rule out the possibility of a person storing all these levels. In the case of subjective discrimination it is usually accepted that seven levels are usually possible. Let us suppose that ρ, v have m, n discriminating levels respectively. Suppose the following axiom is true.

Axiom 1.
    If $\rho > \rho'$, then $\phi(\rho, v) < \phi(\rho', v)$ for all v
    If $v > v'$, then $\phi(\rho, v) > \phi(\rho, v')$ for all ρ
    Then we have at least $(m + n - 1)$ different levels for ϕ.
    Let us add the following axiom.

Axiom 2.
    If $\phi > \phi'$ then $\chi(\phi, \phi'') > \chi(\phi', \phi'')$ for all ϕ''
    If $\phi > \phi'$ then $\chi(\phi'', \phi) > \chi(\phi'', \phi')$ for all ϕ''

---

[1] We discuss the problem of discrimination more fully later on. The point made here is that if high discrimination exists or is desired then the use of ranking has no place in general theories of choice because they cannot be handled subjectively.

Then χ has at least $(2m + 2n - 3)$ levels. Unless our discrimination for ρ, v is very crude, then χ may have a large number of discriminating levels.[1]

The measure with the smallest ambiguity content is Ackoff's 'restricted measure'. Ratio measures are obtained by taking the ratio of two counts. Thus frequency probability is the ratio of the number of times event E happens to the total number of observations. Since there is a limit to the number of observations possible, in any given time interval, then there is a minimal 'distance' between any two non-equal primary probabilities, equal to N^{-2} where N is the maximum number of observations. For derived probabilities the number of levels increases (see our analogous example in the last paragraph), but due to the finite boundedness of any derivation process, there is still a difference below which we cannot measure.

An advantage of measuring by comparison with standards is that to measure $x \, \varepsilon \, \chi$ does not require consideration of all other $x' \, \varepsilon \, \chi$, and hence allows fine discrimination.

The procedure is to choose a finite set of standards $x_1, x_2, x_3, \ldots x_k$, depending on the discrimination level required, and such that

$$x_{n+1} \equiv \lambda^{-1} x_n, \quad n = 2 \ldots k - 1.$$

Addition is normal vector space addition. $z \equiv a \ominus b$ is defined by $a \equiv z \oplus b$. Then x is compared with x_1 to give

$$x \equiv m_1 x_1 \oplus x', \quad x' \equiv x \ominus m_1 x_1$$

Then x' is compared with x_2 to give

$$x' \equiv m_2 x_2 \oplus x'', \quad x'' = x' \ominus m_2 x_2$$

This is repeated until we obtain

$$x \equiv m_1 x_1 \oplus m_2 x_2 + \ldots \oplus m_k x_k \oplus \varepsilon$$

$$\equiv \left(\sum_{s=1}^{k} \lambda^{k-s} m_s \right) x_k + \varepsilon$$

where $\varepsilon < x_k$.

The finest measure, in this case, is then given by $m(x) = \left(\sum_{s=1}^{k} \lambda^{k-s} m_s \right)$.

---

[1] Actually, bearing in mind our remarks in Chapter (2) Section (2.2) and equation (61), v and χ should have the same 'number' of discriminating levels, and, unless, fortuitously $m = n = 1$ this cannot hold except when m, n are infinite. This actually reinforces the cardinality requirement.

It has been argued by Shackle that one of the difficulties of using the probabilistic approach is that to get the probability of E, we need to consider alternative events which may be large in number. This need not be so. Providing we have standard events with specific probabilities, and the requisite axioms of choice, we simply find a standards events between which and E the person is indifferent and we automatically have our measure.

Value also has a standardizing procedure. In von Neumann's theory the probability mixture of two datum prospects is varied until the person is indifferent between this prospect and the one he is trying to value, and the probability obtained is equal to the value of the element being considered.

It is to be noted that the use of the 'indifference' concept is essential in these measures.

Earlier in the section we allowed for the possibility of 'irrational measures' and this needs clarifying. Irrational numbers are, in pure mathematics, simply the limiting points of a sequence of rational numbers. Thus the sequence of rational numbers given by equation (6) has, a limit, an irrational number e.

$$u_n = u_{n-1} + (n!)^{-1}. \tag{6}$$

However, in practical problem solving, infinite sequences cannot arise. Nevertheless irrational measures can arise when determining the optimum solution to a problem. Thus the optimal economic lot size problem has the solution given by equation (7) and this may be irrational.

$$x = (\lambda Q)^{\frac{1}{2}} \tag{7}$$

This is solely a matter of approximation to the corresponding discrete form. We cannot, in complex problems, do without the concept of irrationality, since it may considerably simplify the analysis, which, if carried out in discrete terms would be impossible in many cases.[1]

We have been concerned, in this section with the fact that, effectively, all measures are homomorphisms from a set into the set of integers, rationals, or even irrationals. Depending on the form of measure, even if the meaning of that measure is given, there is still ambiguity in such measures, for the purposes of certain problems.

---

[1] The differential calculus would not exist without the extension (even if only in theory) of discrete variables to continuous variables.

This may arise either because the derivation of the measure is dependent on a subset S of χ, or because of physical limitations on the degree of discrimination ever possible. The selection of the number of discriminant levels is an important problem in the theory of decision and is now given further consideration.

3.3.3 Discriminating Levels

Several remarks have already been made in this context. Thus in situations with several components, certain reasonable axioms increase the number of discriminatory levels possible, and this indicates that objective cardinal measures are more appropriate than subjective ordinal (ranking) numbers. An ultimate extension to include irrational numbers makes the mathematics more tractable, and apparently increases the discriminatory power, although, since all measures are discrete, there is no effective increase in discriminatory power.

Increasing the number of discriminatory levels would allow a higher content of decidability. Why should we want to discriminate between two elements, whose measures are, within the discriminatory powers available at that moment, not discriminated?[1] The answer to this can only be determined by assessing the consequences of using and obtaining such finer discrimination. Let us assume that finer discrimination is, in principle, valid. Let us assume that we have a single value measure θ for alternatives a, b, and that all we know initially is the following.

$$\alpha \leqslant \theta(a), \quad \theta(b) \leqslant \beta. \tag{8}$$

Let us now consider a set of alternative acts, $(d(n))$ where '$d(n)$' is the act 'to add a further n levels of discrimination to $(\alpha \leqslant \theta \leqslant \beta)$'. We shall assume equality of spacing. Let us assume that $c(n)$ is the cost of doing so. $(c(0) = 0)$. For simplicity let $n = 0$, 1 only. For $n = 1$ the possible outcomes are that $\theta(a)$, $\theta(b)$ are in the same or different intervals. The 4 outcomes are as follows, denoted by j. We assume that we limit ourselves to statements of the kind '$z \geqslant y$'.

---

[1] It is unwise to be unnecessarily refined when trying to get more accurate data when the present data will suffice. A common practice is to proceed with crude data and to get more if it fails to produce a decision. When Q contains a small number of discrete alternatives this is permissible, and often results in a decision without refined data.

j: 1.   2.

$\alpha \leqslant \theta(a), \theta(b) \leqslant (\alpha + \beta)/2$ $(\alpha + \beta)/2 \leqslant \theta(a), \theta(b) \leqslant \beta$

j: 3.   4.

$\alpha \leqslant \theta(a) \leqslant (\alpha + \beta)/2 \leqslant \theta(b) \leqslant \beta$ $\alpha \leqslant \theta(b) \leqslant (\alpha + \beta)/2 \leqslant \theta(a) \leqslant \beta$

We then see that the acts a, b which would be taken for each j and n are given as follows[1].

j

		1	2	3	4
n	0	a or b	a or b	a or b	a or b
	1	a or b	a or b	b	a

(9)

The choice of n now depends on which choice criterion one uses. Let $n = 0$. For a maximiner, assuming costs are subtractable from the θ's, the worst that can happen is an outcome α. Let $n = 1$. The worst that can happen is an outcome α and hence a maximiner would not look for further discrimination. A different choice might be made by a minimax-regretter. The use of expectations is not possible as long as ambiguity arises, unless we wish to combine probabilistic and the maximin approach. The probabilities will refer to states $j = 1, 2, 3, 4$, and we will have to take the lowest number for each range.

Thus the discriminatory power desired is no less a problem to be analyzed within our theory than the problems for which it is desired. The desirability will depend on the problems for which it is being considered. If a standard discriminatory power is to be used, independent of the problem, then the appropriate discriminatory power must be determined against the background of a whole class of problems.

Our final section will now be concerned with the derivation of one measure from others, since this constitutes an essential service to much of decision theory.

3.3.4 Measure Structures

It was pointed out in Chapter (2) that the manner in which value measures can be combined to produce other value measures is of

---

[1] In case 3 we may still have $\theta(a) = \theta(b) = (\alpha + \beta)/2$ but we may not know this. We lose nothing by choosing b however. Similar remarks apply to case 4.

considerable importance in decision since it allows a relatively large number of values to be determined from a few using certain rules of combination.

Such operational rules correspond to combinatorial operations in the element space itself in a certain way, and are derivable from such.

To emphasize the nature of the operations with which we are concerned, when measuring something which is a composite of two or more elements, it is worth looking at a misunderstanding which arises in the literature on value theory. There has been a certain preoccupation with the idea of 'strength of preferences'. Such a strength of preference is equivalent to a statement of the form 'A is preferred to B more than C is preferred to D', where A, B, C, D are elements in some space χ. This is not operational and has no meaning, without a prior derivation of values, or without introducing the idea of subtraction in the space χ. What is usually meant is that $m(A) - m(B) > m(C) - m(D)$. Since, in general, m is not uniquely defined, unless some axiom about the measure m and not about the preference on χ is assumed the original statement is not observable unless m is defined and, even then, it may be true for one value function and false for another.

In cases where we are trying to ascertain our measure the proposition 'A is preferred to B more than B is preferred to C', is meaningless. However, sometimes it can be given meaning if a binary operation, '$\ominus$', over χ, exists such that '$A,B \, \varepsilon \, \chi$' $\rightarrow$ '$A \ominus B \, \varepsilon \, \chi$'. This is the basis of all interval measurement. Let $\oplus$ be the vector addition over χ. Let $AI(B \oplus U)$ and $CI(D \oplus V)$.[1] Then we can define $A \ominus B \equiv U$, $C \ominus D \equiv V$. The proposition given initially then has an interpretation that $m(U) > m(V)$. This approach is only valid in the case of additive value functions for which $m(X + Y) = m(X) + m(Y)$, for if $AI(B \oplus U)I(B \oplus V)$ we need $m(U) = m(V)$.

The relationship between operations on χ and arithmetical operations on $m(\chi)$ may take several forms dependent on χ and m. If, for example, χ consists of one basic element and its multiples[2] and if $m(x \oplus y) = m(x) + m(y)$, we will have $m(x) = km(u)$. If χ is von Neumann's prospect space, then the operation combining $u,v \, \varepsilon \, \chi$, will depend on p and, we get $m(upv) = pm(u) + (1 - p)m(v)$. If χ is a

---

[1] I means indifferent to.
[2] (i.e. '$x \, \varepsilon \, \chi$' $\rightarrow$ '$x = u \oplus u \oplus u \oplus \ldots \oplus u$, k times for some integer k').

product space of events, then if m is the probability measure, we have $m(e_1 \otimes e_2) = m_1(e_1) \times m_2(e_2)$.

These are all common simple operations which we have discussed. We can produce more complex ones. Suppose we try to measure the ability of a group of people. Can this be related to individual abilities? If a, b are people εA, if $\chi = A \times A$, and if θ is an operation specifying how a and b will combine together, is there an operation ϕ in $m(A)$ such that the following equation is satisfied?

$$m(a\theta b) = m_1(a)\phi m_1(b). \tag{10}$$

If a, b are the two possible outcomes of an act, represented by $a\theta b$, then, for a maximiner we would have

$$m(a)\phi m(b) \equiv \min \left[m(a), m(b) \right] \tag{11}$$

A more general combinatorial problem arises, in the case of a composite element such as an applicant for a job, in which a, b, c may represent ability, experience, qualifications respectively. A general function of $m_1(a)$, $m_2(b)$, $m_3(c)$ is then needed to measure the value of such a person to the company. Thus $m(a, b, c \ldots)$ satisfies

$$m(a, b, c) = \phi(m_1(a), m_2(b), m_3(c)). \tag{12}$$

An interesting problem arises when overall company performance is related to individual departmental performance. It is sometimes required that, if the company is split into n departments, there exist measures of performance m_i for the ith department such that, for some ϕ, company performance satisfies the following equation.[1]

$$m = \phi(m_1, m_2, m_3 \ldots m_n) \tag{13}$$

If equation (13) does exist then each department simply tries to maximize its own measure to maximize company performance. Thus sales performance might be measured by the volume of demand in any given time period, production performance by average delay in satisfying orders, engineering department by the cost of maintaining a certain level of service. These examples are not very realistic, since cost must enter into sales and production performance, and customer satisfaction may depend on delays and quality. However, there is evidence that the principle is valid[2] and there is a definite advantage

---

[1] This gives formal expression to the notion 'management by objectives'.

[2] Naturally, of course, this gives rise to suboptimization but can still improve performance.

in this approach since otherwise the whole organization would be involved in everyone elses decisions, giving rise to an impracticable decision organization.

Thus we see that measurability is of fundamental importance in decision theory. The previous notes do no more than shed some light on its nature and of the problems involved in measuring and using such measures. It must not be assumed that measuring with as fine a discrimination as possible is the objective of any application of decision theory to problematic situations. As with the point made about the illusion of the quest for certainty, so for all measures, must we consider such problems from the same point of view as the problems for which they are derived. Apart from time and other limitations preventing the attainment of such fineness of discrimination, there are other consequences of an economic nature that put definite boundaries on the desirability of such. We will always have some residual ambiguity in empirical analyses, and the appropriate amount generates a problem itself.

Summary

So far we have discussed the main aspects of modern Decision Theory, viz. value and uncertainty in the context of choice. However decidability enters into problem solving through other avenues than those cited above. It is, for example, just as much a decision if we can decide the probability of a compound event from knowledge of the component probabilities, providing we stick to our general concept of mathematical, or logical, decision; this arises because, prior to the analysis, there are alternative possible probabilities from which we could select, and the analysis effectively selects the appropriate one. In this chapter we pursue the general notion of decidability.

3.1 Complete, Probabilistic and Partial Decidability

The notions of 'proof' and 'decidability' are compared. It is recognized that the answering of any question is a matter of decision. The notion of 'the probability that a statement is true' is reinterpreted in terms of probabilistic choice (see Chapter (2)). The two possible interpretations are that 'the probability, that a_i is the correct choice, is P_i' or 'a_i will be chosen with probability P_i'. The dangers and difficulties of the former interpretation are examined. The conclusion is that neither should we accept answers to questions because they are most likely to be true, nor should we select alternatives because they are most likely the right (or best). Ideally we would wish to recommend a definite alternative but if we accept the ideas of probabilistic decidability we would recommend that the choice be made probabilistically.

The concept of partial decidability is discussed and the effect of this is simply to reduce the size of the problem and the final choice is made subjectively. The well known concept of Pareto Optimality arises here.

It has to be realized that the search for complete decidability is misguided since this can be a costly process.

3.2 Decision and Axioms of Choice

We now consider the main means of achieving decidability in problem solving, viz. the use of the axiomatic approach. Briefly the idea is that, on the basis of the establishment of certain axioms in simple standard situations, we can decide issues of a considerably larger complexity, in which the human is unreliable computationally.

The axioms used in the theories of choice are taken as illustrations. These are constructed in a general language which is then interpreted in the context of special problems areas. Thus von Neumann's axioms are in terms of a general set M, but can be reinterpreted in terms of the variables describing physical decision areas such as investment, personnel selection, etc.

Consideration is given to the various characteristics of such axioms occurring in the literature, with a view to determination of their import for practical decision-making. These are 'topological', 'mixture', 'order' and 'group' axioms, and it is established that the last two are the ones which are of greatest relevance.

It is possible to describe how choice is made in areas such as investment and inventory control, but if we wish to know if they are compatible we need to be able to express them in terms of common variables, e.g. cash and probabilities. Thus the resolution of whether or not the actual choice behaviour is valid can only be made within a more general framework. This is theoretically achievable by tracing through the effects of every event until a common terminal representation of all decisions is achieved. This chain is, however, unlimited and the terms in which the effects of choices are expressed are selected at a subjectively chosen stage. Thus we may express our choices in terms of quality of product, without being able to trace this through to its cash consequences. Our decidability content is therefore limited by this subjective valuation. The Extensive form of decision-making (Chapter (2)) becomes useful. As a result of this process we will evolve decision-making procedures in different areas that may, in fact, be incompatible.

The axiomatic approach will apply to all problems of measurement, as well as the ones discussed, viz. value and uncertainty.

3.3 Measure and Decision

The mere allocation of a number to an element does not constitute measurement. Measurements will have properties and it is these properties which contribute the decidability to our problem solving.

We concern ourselves with the meaning of measure, which can lie in its operational definition or in its use. There is no conflict in these two points of view. The misuse of measures is stressed in typical situations.

In order to describe some situation for certain purposes, several measures may be needed, and there often arises the problem of which ones to make. In theories of choice we postulate that there are two, viz: value and uncertainty. In other circumstances the identification of a minimal set of sufficient measures is not easy.

There are several forms of measure which are in practical use, viz. numbering, counting, ranking and strict measure.

These play various roles in the decidability issue, but have varying degrees of ambiguity content and hence have varying decidability contributions. Ambiguity is tied up with the degree of discrimination and it is easily seen that even in relatively simple theories, such as Shackle's, the number of discriminatory levels is above the commonly accepted discriminatory power of a subjective assessment. Thus, if the theory is correct, it is likely that the restricted measure is the appropriate one. In the theories of value and uncertainty discussed the restricted measure is the dominating one. The point is made that, although irrational numbers do not arise in practice, the restricted measure allows their use as an approximation to reality which facilitates the computational aspects.

The desirability of fine discrimination levels has to be considered in the light of the costs and benefits involved with such efforts. We certainly do not wish to have precise measures if they cost the earth and the gains are small. A simple example is used to illustrate this as a decision problem itself.

Finally we consider the generalization of the value structure and uncertainty structure arguments to the case of general measures. Basically we wish to determine the relationships between measures of composite elements and values of the minor elements. A popular misconception in the use of 'strength of preferences' is used to illustrate the point. A typical problem of finding ways of measuring departmental performance so that company performance can be assessed in terms of these is used to illustrate the objective of the exercise.

CHAPTER 4

Some Practical Considerations in Decision Theory

Chapters (1), (2) and (3) deal with the more fundamental aspects of decision—with its meaning, with the basic theories of uncertainty, value and choice, with decidability, in general. The research has, as its prime purpose, the identification of the relevance of such work to practical problem solving, and we shall devote this chapter to a discussion of some of the practical problems which arise in interpreting and implementing some of the results in practice. This is distinct from the content of Chapter (6), which concerns itself specifically with the senses in which improved decision-making can be achieved.

4.1 Improved 'Selection' versus Improved 'Consequences'

We are concerned primarily with the 'selection and selection process' and only with 'consequences' in as much as they motivate our selection. Decision Theory is partly a logical theory, and is concerned with deciding actions on the basis of various premises, and with the determination of such premises. The scientific content in our research lies in the determination of the premises. It is not to be assumed that it is a general consequence of applying Decision Theory that the consequences of any act or class of acts will be improved. Whereas much scientific work purports to be 'exact' to a large extent, and the search is for this exactness and truth in general, practical problem solving must consider the economic resources which are required to carry out such a search, and hence actions may be taken at points at which a great deal of residual uncertainty exists. The fact that such actions may be decidable does not mean that the person will be better off having decided, although we would not preclude any parts of our theory which enabled us to make such statements. Indeed, there are areas in which the uncertainty, although sometimes not completely removable, is small enough to be effectively ignored.

Chance plays a part in the consequences of any act, and not only must we query the economical ability to reduce it beyond a certain level, or even, as in Chapter (2), query the theoretical possibility of doing so (depending on the nature of the ambiguity), but we must also query the motivations behind the quest for precise identification of consequences. Since our researches are for the benefit of humans, they should be considered as an integral part of them. As such, it is not correct to assume a person wants to know the consequences of an act. Leaving aside his wish to play 'make-believe', he may prefer to be doubtful about the outcomes rather than to know, for certain, what will happen. A great deal of the spur in life arises because 'so and so might happen' (if it is beneficial) or because 'so and so may never occur' (if it would be disastrous). A far more serious possibility arises if we are considering 'enlightening' everyone, for then, if the consequences of every act were predictable,[1] not only may certain individuals suffer, but, dependent on whether a suitable group value existed, the group as a whole might be worse off.

Whether or not a person does want certainty can only be answered by treating it as a problem, analyzable just as any other problem for which we are constructing our theory. Thus consider the problem, given below, in which we have two states of nature, and two acts.

If we knew the true state of nature we would know which act to take. Hence if we introduce a third alternative (viz: to determine the

		state	
		1	2
act	1	0	1
	2	2	0

---

[1] The reason for the 'if' is that consequences are partly a function of acts yet to be taken by the person himself and other people. Thus, if the consequences were predictable, then this choice behaviour is predictable. Since this particular act forms part of the consequences of some future act, the assumption of predictability of consequences implies the predictability of all choice, and hence there would be no prescriptive content in this work.

true state of nature, at a cost c) we have a new problem given below[1].

Whether or not certainty is required in this problem depends on c and on the appropriate choice criterion. Thus a 'maximin' person

state

		1	2
	1	0	1
act	2	2	0
	3	$2-c$	$1-c$

would settle for certainty only if $c \leqslant 1$. A 'maximum expectation' person, who gives probabilities p_1, p_2 to states 1, 2 respectively, would choose certainty only if $c \leqslant p_2$ and $c \leqslant 2p_1$. In both cases, although he may wish to improve the outcome, he realizes that chance enters into it, and hence it poses a problem for him which he resolves according to his own risk characteristics.

There is considerable difficulty in defining 'improved selection'. In Chapter (2) we stressed the fact that choice between two acts a, b, may be relative to the problem Q of which they are components. Unless a is preferred to all $b \, \varepsilon \, Q$ it may make no sense to say '$a \geqslant b$' for any specific b in problem Q. Savage's minimax-regret criterion illustrates this, and, in as much as a person may behave according to this criterion, we cannot talk in terms of 'improved selection'. Such a concept has no intrinsic part in Decision Theory. On the other hand such people may be said to be 'irrational' and dismissed from the domain of enquiry by insisting that a value structure exists permitting the use of the term 'better selection'. This simplifies our task somewhat as was pointed out in Chapter (2).

When considering 'improvements' we have to consider both the selection procedure and the selection itself. The improvement may not necessitate any change in solution to the problem. It may lie simply in replacing the human by a mechanical aid to some extent, and thereby saving some of the time and effort, which he can then

---

[1] We assume the entries in the matrix represent the values of the consequences.

use for other purposes. This again generates a problem the resolution of which depends on his own preference characteristics. He may feel that, even though the ultimate selection is the same, he would, for various reasons, prefer to do it himself. Such reasons may be purely psychological, the wish to maintain personal contact with the problem, or with other people involved, or even because of fears of what such steps may bring in the future. It is his problem and he has to choose, bearing in mind any relevant implications.

Improvement in the actual choice is the commonly accepted viewpoint.

If we can demonstrate that the outcome will be better, we have no problem. To do this we may bring to bear our mathematical powers, and any relevant information, to enlarge the scope of the problem by covering a wider range of alternatives. This applies particularly where human reliability, in deduction alone, is in question.

In cases where uncertainty is a limiting factor we may still, providing we can determine the appropriate criteria and values, enlarge the scope of the problem and produce a 'better choice' in the sense that 'if the person himself could have managed the analysis, he would have chosen the same one'.

In many instances we cannot go so far as to prescribe the selection, and yet we can add some decidability content to the selection by providing information of various kinds. Thus we may be able to present the outcome of an investment policy as a probability distribution of cash flows, although we may have great difficulty in getting a value function and, hence, in recommending an action.

The above paragraphs serve to give a brief idea of the sense in which improvements may be interpreted. The next few sections will be concerned with certain practical points which have to be considered before any conclusive idea of where Decision Theory studies will impinge on practical problem solving. To be able to do so will require consideration of both the actual selection processes, the recommended processes, and, by no means least, the environment within which such selection takes place. The person himself is an integral part of this. We will now consider the nature of the subject matter of our studies, particularly in relation to the physical sciences, which constitute the objective part of our area of investigation. The decision-makers constitute the subjective part. The following diagram gives the basic interplay between the objective and subjective aspects of the system:

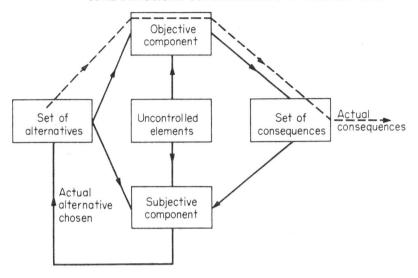

This model applies to the secondary, as well as primary, decision.

4.2 Physical versus Human Sciences

'Operational Research' is often portrayed as bringing 'scientific method' to problem solving. It is considered to be to some extent a valueless occupation to press for definitions of this, but, in as much as it is the present attitude that 'science is needed in business operations', we must give it some consideration, especially if the same degree of exactness is sought after in business studies as has been achieved in the physical sciences.

Smith (64) identifies 'scientific method' with a process of 'prediction'. The end product is the prediction of certain outcomes under certain conditions. It is implicit in the search for certainty propositions about behaviour of any kind. In macroscopic physical sciences (i.e. in the application of scientific method to the domain of macroscopic physical entities) a degree of exactness has been achieved which puts uncertainty below a minimal threshold. The domain of the business world contains physical entities and human entities whose characteristic behaviours form the nucleus of any recommendations in relationship to problem resolution. Thus many problems, which are of concern to us, are purely technical ones

117

arising out of the determination of the appropriate manner to run (or design), say, a multistage production process, and, subject to the existence of sufficient criteria of performance, their resolution depends on physical theories. Thus chemical reactions depend, in a relatively exact way[1], on environmental temperatures and pressures, which can be controlled. The human aspect arises in the determination of the 'appropriate criteria' and this must stem from the concept of 'preference' and ultimately be based on observed choice in problematic situations of various kinds. The determination of such criteria is subject to the 'scientific method', in principle, since it follows the same set procedure.[2] One can rightly say that the foundations of decision theory lie in the predictability of choice, and of physical behaviour, under specified conditions.[3] The decidability content enters into it only in as much as certain premises concerning behaviour can be used to determine the outcome of an action, or the choice to be made, in a specific problematic situation.

One very real difficulty now presents itself when we consider the rejection or acceptance of the objective and subjective theoretical aspects of any analysis. The truth or falsity of any theory is only important in as much as its acceptance or rejection has consequences for the future. It is therefore, itself, subject to choice. Such a choice is equivalent to asserting the intention 'to act as if the theory were true or false'. We would therefore require a theory of choice with regard to problematic situations in which the alternative acts were 'to accept a specific theory', and, although one must bear this in mind, we cannot pursue it here.

Typical of this is the problem of using incorrect criteria in a given class of problems. Contrary to the usual approach, it is of no validity to test the robustness of solution to each set of criteria. Only if the solution is invariant under the criteria is this meaningful. Thus, if we are in doubt as to whether the Laplace criterion or the Wald criterion is the one to use (errors being attributed to interference) in an inventory control problem, the nearness of respective solutions is no guide, logically speaking, to removing our dilemma.

Strictly speaking we should evaluate the consequences of assuming

---

[1] With small degrees of uncertainty.

[2] 'Conditions' include the 'problematic class' on which the observations are being made.

[3] Observation—formulation of theory—predict choice in given problematic situation—observation—amendment of theory . . .

that Wald's criterion (Laplace's criterion) is true when Laplace's criterion (Wald's criterion) is true. The difficulty with this approach is that the consequence values of an alternative are not comparable in different situations. Thus consider the following primary problem.

$Q_1 \equiv acts$	states j		
	1	2	
1	4	0	← Laplace Solution : L
2	1	1	← Wald Solution : W

The act values, under the different criteria, L, W, are as follows.

acts	criteria	
	L	W
1	2	0
2	1	1

The choice of L or W then gives rise to the following problem Q_2.

$Q_2 \equiv acts$	states	
	L	W
L	2	0
W	1	1
	Laplace act values	Wald act values

The consequences in column L are not comparable with the consequences in column W. It is not enough to replace an act by an equivalent one with a fixed certain outcomes. Thus if we replace

119

act 1 in Q_1 by (2, 2) this would then become a Waldean solution as well and Q_2 would become

	states		
	L	W
$Q_2 \equiv acts$ L	2	2
W	2	2

We would then be indifferent between L and W.

In a case where the Hurwicz α criterion is taken as valid, but α is unknown, we have a common value system for acts whatever the true or chosen value of α happens to be and hence there is a consistent choice procedure for the counterpart of Q_2.[1]

If a theory is not in conflict with present observational knowledge, it does not mean it is true. Neither does the fact that it conflicts with observational knowledge mean that it is false, since errors will be made in recording observations or the results might be distorted because the operations involved in verification actually change the conditions we may think are fixed. If the 'errors' are small, they usually are attributed to one of these two reasons and the theory is accepted, provisionally, irrespective of the nature of any other problems for which it provides a premise. Good (32) advocates a 'Theory of Conclusions' for such cases. This is is generally the case with physical sciences.

We could extend 'truth' or 'falsity' of any theory to the 'probability of being true or false', but if, instead of having to choose a theory as being true or false, we were asked to assess the probabilities that we would choose it to be true or false, we would face considerable difficulties. Thus suppose we considered Luce's probabilistic choice, and observed that, in the problematic situation given by $Q \equiv \begin{bmatrix} h \\ \sim h \end{bmatrix}$, h was chosen 40 per cent of the time and h was chosen 60 per cent of the time, would we use this theory? The complicating issue is that

---

[1] In carrying out experiments it may be that no specific α will give perfect prediction. It may be thought, however, that the lack of perfect prediction may be due to interference and not due to the inaccuracy of the theory. Even if it were predictive for the experiments there is no guarantee that the same would apply in more realistic situations.

of deriving consequences of incorrect choice. It may seem reasonable when the choice is in terms of physical alternatives, but it is inconceivable that it could be the case for theory type alternatives. We are likely to have to take the first interpretation of '$p(h) = p$' discussed in Chapter (3), Section (1), with no consideration of consequences of any action dependent on the truth or falsity of h.

There is a difference between theories which make propositions about observable events, whose observation confirms or refutes the theory, and those theories which make propositions about some probabilities, which cannot be confirmed or refuted. If we are concerned with the concept of limiting frequency probability, which forms the basis of all scientific inferences at the moment, the acceptance or rejection of such theories must be dependent on the person concerned as are reflected by the consequences for him personally. We do, even in the physical theories, reach a state in which the theories are probabilistically oriented. This occurs in the area of Quantum Mechanics whereby attempts to measure individual behaviour of electrons is supplanted by measuring macroscopic behaviour of ensembles of such elements, each attributed with a probabilistic behaviour. The theories which are verifiable are then the observable macroscopic theories. Since, in such theories, it is the macroscopic behaviour which has utility, in that it is the properties of ensembles which are used in practice, we have no difficulty here, but in the Decision Theory area it is the individual to whom such theories are directed at the moment, and it is the individual whose behaviour we want to determine.

We thus see that our theories are either predictive, non-probabilistic and observable, or predictive, in terms of probabilistic propositions, and not observable (in the sense of confirmable or refutable). The latter type are very much subjectively oriented.

In as much as we are dealing with individual elements we must take note of Heisenberg's Principle of Uncertainty (Chapter (2), Section (3)).

It has been well established that a person's behaviour can be considerably affected by artificial experimental set-ups which are removed, to some extent, from the problem environment in which they are eventually going to operate. Even the presence of an observer can influence their behaviour. The fact that von Neumann's theory may hold for a given experiment does not mean that the same 'values' would pertain in an unobserved similar problematic environment. Thus the essence of 'fun' or 'novelty' in the experiment may arise. Even an effort to 'fox' the experimenter may arise. As Ackoff (3) points out, the faculty of memory possessed by the human can induce a certain consistency of choice in a small time-span experiment, which is not there in longer life cycles, these being permeated by a heterogeneous collection of miscellaneous events and problems. Thus, because he chose *a* rather than *b* to begin with, he might always choose *a* rather than *b*. Transitivity would no doubt exist because of a wish to appear 'rational'. One might say that this 'artificial experiment' is more real than the real set up by virtue of the fact that it is uninterrupted, and that consistency ought to exist, but this problem remains to be resolved.

Perhaps the essential difference between theories of physical behaviour and theories of human choice behaviour is the ability of the latter to 'think'. Knight (41) expressed it quite well in:

'For all we see or for all science can ever tell us we might as well be unconscious automata, but we are not. ... We perceive the world before we react to it and we react not to what we perceive, but to what we infer. ... We must infer what the further situation would have been without interference and what change will be wrought by action. ... The fundamental difference in the case of animal or conscious (as from vegetable and unconscious) life is that it can react to a situation before it materializes; it can see things coming.'

It is the very cognitive ability of humans that influences the ability to develop predictive theories of choice. This arises primarily because the outcome of any act depends on the future behaviour of many other human beings, which will be a function of conditions existing at the time. As a consequence of the implementation of Decision Theory, which brings to light certain knowledge about the consequences of an act, some behaviour, as a function of certain conditions, will change. Thus, as Knight points out, a man's power of reasoning coupled with this extra information, allows him to select a 'better' alternative.

The determination of the problematic class and condition specification in terms of which our basic theories of choice are to be expressed poses a severe difficulty at this time. We can have theories of choice in terms of specific functional problems, such as personnel selection, investment selection, or inventory policies selection. We can have theories of choice expressed in terms of general problems based on ultimate indivisible consequences. Thus a specific functional theory would state that the criterion

$$\sum_{i=1}^{n} \alpha_i x_i$$

was used to select personnel, where $[x_i]$ were measures of certain characteristics. A general theory would state that if we could relate the consequences of $[x_1, x_2, \ldots x_n]$ to, say, a probability distribution over money, then by using von Neumann's theory to value this probability distribution of money we could decide what the criterion function ought to have been. Thus the general theory enables all the functional problems to be resolved.

In functional problems one is faced with the difficulty of identifying the relevant conditions which the problem solver observes. Thus in personnel selection one may not be able to determine all the conditions considered in the selection at that time. Some will be attributes of a person which are not expressible formally. Some will relate to the company's needs at that time. All are likely to have a bearing on the suitability of the candidate for the post.

The use of a general theory is desirable but only useful if we can relate the conditions specifying the problem, and each specific act, to the consequences in terms of which the general theory is expressed.

This is not very often possible and it may be that either for the reason of just mechanizing the choice behaviour, or for the reason of communicating the apparent emphasis on various factors to other people, or for the purpose of constructing value functions for the purpose of extending the analysis of complex problems having sub-problems of this kind as components,[1] we will accept such theories of choice as useful and not wish to try to change behaviour in this class of problems. In such cases we can expect to find instability by virtue of the fact that the cognitive aspect may give rise to different inferences for apparently identical conditions, perhaps because of inconsistent reasoning, or perhaps because of added reasoning. Thus the importance of certain attributes may change as time goes on. Here again the cognitive aspect plays an important part and renders certain behaviours likely to be unstable, something not present in physical sciences in which there is no effect of past history on future behaviour.[2] One might overcome this by including all past

---

[1] See Ackoff (2) and Chapter (2) Section (2.2.3) of this book for such uses.

[2] Thus the true laws of electron behaviour as a function of conditions will be the same in future as now.

history in the condition language, but we would then have no repetition and hence no theory.

The difficulties of relating conditions to certain basic consequences, therefore, give rise to the need for a subjective functional theory of choice in terms of certain conditions. However the construction of such theories suffers from several deficits which do not arise in the physical sciences. In the first case our ability to select our actions contributes towards a generation of conditions previously unencountered whose implications for the future are not assessable. Only if we had a closed system which regenerated itself every so often could we construct an anywhere near exact theory.[1] Secondly, the events which mature at any point in time may depend on a succession of conditions. Thus we might have (and will have) autocorrelation of quite high orders. Thirdly, conditions are a function of past choices just as much as they influence future choices and therefore, if one is thinking, by reason of introducing a decidability element into choice behaviour, of changing some choice behaviours in certain problematic situations, the inferences based on any condition existing now need not be consistent with the events following the 'same' condition at some previous time. This two-way dynamic causal relation will change with our choice behaviour. A study of the method of analyzing a class of problems relating to district office organization in an insurance company (White (73)) showed this up quite convincingly to such an extent that little relationship between any of the control variables and suitably chosen consequence variables existed. Conditions, economic and sociological, were changing continually, and past performance was a function of a wide variety of other actions previously taken, not now in the class of actions to be considered. The conditions were not regenerating themselves. No analogous problem exists in the physical sciences since the element of 'cognitive choice' is missing.

It is possible, in the physical sciences, to isolate the problem on hand in a self-contained experiment. Within the field of human behaviour this constitutes one of the severest of difficulties to a feasible practical experimental set-up.

The sequential nature of all our problems has already been stressed in Chapter (2). Any bounded experiment must suffer, to a greater or less degree, from an unsatisfactory 'cut off' of the problem

[1] By this we mean that the system only moves within a bounded, or finite, set of states.

from the whole sphere of its influence, which extends over some time period and is interdependent with other problems yet to be solved in the future.

Consider an effort to verify von Neumann's theory, in the case of problems generated over monetary returns. Each individual experiment is supposed to be isolated from every other individual experiment. However, the effective consequence is likely to be measurable by overall performance: therefore the real problem is then one of choosing a strategy for the sequence of problems depending on what is known about the number of 'experiments'. Unless we are prepared to assume that these individual 'experiments' do constitute isolated non-interacting problems, we only have one experimental observation (a complex one no doubt, but, nonetheless, only one). The difficulty arises only if we are going to reward someone on the basis of the outcome at each stage. Another alternative is to ask what the choice would be, and this suffers severely from the artifiality of the experimental set up. Quite obviously we will have to put up with this latter approach when large quantities of money are involved.

If we now try to set up 'policy' type experiments,[1] we will fall into the same difficulty, since a sequence of sequences is still one sequence.

A further difficulty arises from the fact that the consequences of any finite sum of money, accumulated at the end of the series of experiments, may not be known, since it is not to be assumed that money is indecomposable. It has its uses, and, unless we hoard it for its own sake, these are relevant to its value. Therefore, although we may determine values, these may not be the valid ones. After all, as Smith (64) says, 'values only describe choice in given circumstances'[2]. They have no absolute connotation.

Considerations of this kind do not arise in physical theories where the ability to repeat isolated experiments is the main requirement for verifying such theories. In theories of choice the independence of experiments in the series has to be demonstrated somehow.

We thus see that that subjective part of our subject matter, which is connected directly with theories of choice, is essentially a much more complex area than that of the physical sciences. The main difficulties arise from the complexity and variety of motivating conditions, from their dependence on choice, from the effect of the cognitive ability of humans, and from the difficulties in isolating

[1] i.e. those in which the alternatives are policies operating over a given time period.
[2] This is the main theme of Chapter (2) Section (2).

problems either for experimental purposes or for solution in practice. There will be an obvious limit to the content of decidability we can bring into problem solving on these counts. We will always have difficulty, in as much we will never be quite sure of our premises in 'verifying' recommendations made on the basis of any theory we put forward. It is the task of the theory to search out the more valid areas in which some decision can be reasonably introduced.

It is likely to be true that, except in several relatively well-defined low dimensional sequential problems, a lot of the useful and demonstrable theories will be those relating to classes of problems which can be isolated and contained within computable bounds. Problems not able to be isolated in principle may be made so at the risk that subjective value assessments of decomposable consequences (in principle) are in some sense invalid. Thus, in problems which are a composite of investments in research and personnel, the problems can be isolated if we are prepared to consider such elements as indecomposable and set up choice experiments in terms of these directly. Quite obviously subjective judgements are essential for any computable theory covering complex connected sets of problems.

Any satisfactory pragmatic theory of decision must pay some attention to the problem environment, for, after all, we are not concerned with improving isolated problem resolution, but rather with improving the resolution of our aggregate problems, over which we have to expend our resources.

We must also look at the problem-solving behaviours which exist, for it is likely to be the case that a sensible derivation of improved problem solving will need to be developed from knowledge of existing procedures.

4.3 Problems and Problem Solving Behaviour

Along with the theoretical surveys carried out, consideration was given to the variety of problems encountered in business. Such variety stems from the different roles which various individuals play in a company's activities and also from the need for an individual to decide on many issues within the role defined for him.

Selection processes can be 'nested', i.e. a selection process may involve several forms of ambiguity[1] which have to be resolved in a

[1] The resolution of these ambiguities is often referred to as the 'methodology of problem solving'.

definite order before the primary selection can be made. For example we might have the following process:

Primary Problem Q_1: which act $a \, \varepsilon \, A$ should be selected?
Secondary Problem Q_2: which data $d \, \varepsilon \, D$ should be obtained?
Secondary Problem Q_3: which method of analysis $m \, \varepsilon \, M$ should be used?
Secondary Problem Q_4: should we return to Q_1, Q_2 or Q_3?

The actual process will be as follows, where s, s' are states of knowledge existing between problems.

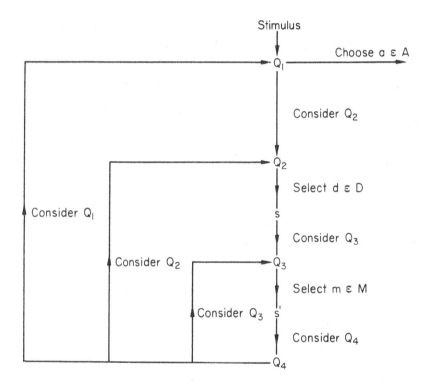

This is quite a simple illustration, and things can get a great deal more complicated. Q_1 may itself be unclear. Thus a communication requiring action in a certain area may leave the decision-maker with some ambiguity as to what sort of action is required or even why the action is required. This generates another problem. He can either continue under his own steam making his own presuppositions about these points, or he can go back to the person who made the request. This activity can, itself, have several cycles. Another possibility arises, again concerning Q_1, in that a stage may be reached in which no $a \varepsilon A$ is thought to be satisfactory, and then the problem of whether or not to enlarge A is generated.

This applies to every problem Q, which must initially have its domain of alternatives defined at least in as much as, even by some implicit means, any a can be classified as in A or in $\sim A$.[1]

Such secondary problems are likely to be, to a large extent, resolvable by choice alone, in the absence of criteria. An investigation of the problem solving process in a complex problem at an insurance company (see (73)) has provided evidence on this, particularly in the use of mathematical models. It has served to stress the point that such problems are as integral a part of selection behaviour as the primary problems and should be treated as such. Not only are time, effort and money devoted to such analyses, but they are, as with any other selection problem, subject to risk in that the worth of the results is not initially known.[2] It has been emphasized in some of our more mathematical studies (forthcoming Ph.D. thesis J. M. Norman (54))[3] that the model and solution procedure used in any analysis has to be selected against the background of the primary problem. There is a two-way effect, viz: the solution to the primary problem depends on the results of the mathematical analysis, and, in as much as these results depend on model used, these results depend on some information about the primary problem. If the primary problem involves a high monetary level, then the selection

---

[1] Thus we may allow for A to be determined, not by extensional form, but by some rule, e.g. if $c(a)$ is some monetary cost of a, then if $A = \{ a : c(a) < \alpha \}$ we need not identify A initially, but can test each a at an appropriate stage.

[2] It is interesting to note that if we accept the idea of probabilistic choice, then we ought to accept this for all problems. Thus given various alternative mathematical models we ought to allow for the possibility of randomizing over these.

[3] One of the basic difficulties in using Dynamic Programming is the extensive computational time needed. However, resorting to the use of simulation severely limits the alternatives which can be considered.

of the model and solution mechanism may be made on economic grounds, e.g. if the order of money involved in Q_1 is £100,000 and then any $m \varepsilon M$ (where M is a set of possible methods of analysis) such that $c(m) \leqslant$ £500 might be acceptable. Thus, given this information about Q_1, m might be partially decidable.

Having exhibited the complexity of, and variety in, even the simplest primary problems, it has to be realized that, depending both on a person's function in the organization, and on his wish or ability to ask questions, and hence to generate more problems, he is operating in an environment in which problems are arriving according to some time pattern, partly controllable by himself, and partly outside his control. Not only do the selections for one problem impinge on other problems, existing now and in the future, but the time, effort and money used in considering any one problem reduces the same quantities available for other problems. Thus the 'super-problem' of how to analyze each problem in this complex stream is essentially quite important, although it is, at the moment, essentially a matter of choice.

It has been a criticism of analytic approaches that they are of little help in 'large problems' and are not needed in 'small problems'. This deserves investigation, but it has to be borne in mind that the outcome of an act in any large investment problem depends on a variety of other problem solutions, of varying 'sizes', within the organization. Thus one 'right' large scale selection can be offset by many 'wrong' small scale selections.

The manner in which problems are resolved, as they arise, depends on the nature of the problem. However, it is not to be accepted that this manner, whatever it is, is to remain unquestioned. The procedures will vary from use of complete decision rules[1] to one of pure choice.[2]

At the lower levels of operation in an enterprise, there is usually very little scope for personal choice, and the functional problems are not only repetitive, but do not differ a great deal in variety. Thus there are rules for determining how much tax to deduct from a person's salary; there are rules for determining premiums for certain

[1] Or 'programmed decision-making' as Simon (63) terms it.

[2] If, in any process, any component problem is decidable, the process is partly programmed for the process and completely programmed for the component problem; thus, even though the choice in Q is not programmed, the selection of, say, statistical models might be, once it has been accepted that a statistical analysis is to be carried out.

129

policies in insurance; there are rules for determining process settings for various qualities of product.

If it were not for the fact that a high proportion of problematic situations are decidable formally, on the basis of such rules, not only would time, effort and money, as resources needed to resolve such problems, be considerable, but indirect costs of delays and errors would arise. However, not only does this generate the problem of the appropriate choice of such rules, but it also generates the problem as to whether they should be complete to any specific extent.[1] Thus, although complete rules remove most of the mental effort and time they do have their attendant disadvantages. In inventory control, complete rules may not allow for changes in environmental factors assumed not to take place when deriving such rules. Assumptions of statistical stability of demand are upset by slumps. Such situations operate by rules which are themselves complete at any one time, in terms of a specified set of conditions, but subject to change by some monitoring controller of other economic conditions. Attempts to complete them formally, by building in economic conditions, usually fail because of the controller's faith in his ability to make inferences about the future which are better than any statistical analysis, the latter being limited because of lack of data and the generation of new conditions.[2] This limited completeness applies to many areas where the benefits of further analysis are deemed less than the benefits of subjective control at suitable points in time.

One aspect of decision-making common to all levels of an organization, which centres around the degree of completeness of decision rules, is that of 'delegation of decision-making'. This problem was a prominent one in the studies carried out in White (74). In that paper a variety of complete or incomplete decision rules in an insurance company are discussed.

Delegation of a problem area by one person to a subordinate may

---

[1] In principle, because of the inability to parametrize the environment by a finite set of parameters, every decision rule, depending only on a finite number of parameters, as it must do in practice, cannot be complete for everyone, although it may be effectively complete for periods of time. Thus, if the conditions, not within the language of the decision rule, vary very slowly, we may assume them fixed for a certain length of time.

[2] This is discussed in Chapter (4) Section (2). Thus the demand on a workshop may depend on the knowledge that certain changes in the nature of components used are iminent and forecasting techniques based simply on past history of demand are risky.

involve the transfer of complete authority for ultimate decision within that class of problems. On the other hand they may not be completely transferred, since the rules governing the transfer may allow reference to him in cases of doubt or in special cases. The gains may be that someone else may be able to make a better decision than the person himself, if they have more time. Alternatively he may just gain the time he would otherwise have spent on resolving them. The losses may be that the delegate may not be as competent as himself, that the delegate may have different values to himself, or, even if he were not to act according to these, he may not know his superior's values.

Apart from the problem classes to which some degree of programming (decidability) may apply, there are the less frequent problematic situations for which, for various reasons, the possibility of programming appears to be untenable. These problems, although frequent in totality, form a heterogeneous class, the subclasses of which contain infrequent occurrences. Thus investment problems, merger problems, sales strategy problems, personnel policy problems and customer relation problems, arise from time to time but, because the groups appear to be different, the group members occur infrequently enough not to be feasible for programming. However, although they are not programmed, in as much as a definite decision rule giving a one-one correspondence between the problem and the selection exists, they may be programmable in the sense that 'a problem solving procedure' now replaces the usual decision rules.[1] Since such problems are viewed in the light of their effects on company performance they have a common denominator. Such a problem resolving procedure, appropriate to all problems, would form a 'meta decision rule' for all our problems. Let us look at Ackoff's problem solving rule as stated in (2). This forms the foundation stone of Operational Research, which is concerned with applied problem solving. The rule is implicit in the following steps.

(i) Formulating the problem;
(ii) Constructing the mathematical model to represent the system under study;
(iii) Deriving a solution from the model;

[1] For our purposes 'problem solving' is distinct from 'action-selection'. It is concerned with an analysis of the primary problem, through a series of secondary problems, and not just with selecting an action in a specified problem—this latter element being the main body of Chapter (2).

(iv) Testing the model and solution derived from it;
(v) Establishing controls over the solution.

Although this serves to orient our approach towards the solution of our problem, it is ambiguous itself, and inevitably poses problems for the person concerned. What is meant by formulating the problem? This must include a preliminary specification of alternatives. How do we get these?[1] In addition no allowance is made for the resolution of problems concerning changes in this formulation, which may arise later in the process.[2] Ackoff also includes the determination of 'objectives' in (i). This depends on whether one wishes to prescribe choice or simply to supply information about the outcome of any act. By no means all Operational Research studies go further than the second form of analysis. How do we determine 'objectives'?

In order to be complete stage (i) should specify all problems, $\{Q\}$, arising in the stages (i) to (v), and these can be considerable in number. Similar remarks apply to the operational interpretation of the remaining stages. Constructing a logical connected picture of this is quite a task, but if decidability is required, at a cost, this must be done.

Quite obviously we do need some procedure for problem solving if we want to bring decision, at any level, into the selection. This is no more than a tautology. No one need ever decide, but then he pays the penalty for absence of reasoning, whatever this may be.[3]

Whether one procedure can be said to be better than another may depend on whether its logic avoids pitfalls one wants to avoid. Thus any procedure not including (v) may be deemed to be inadmissible, although the control cost may be prohibitive. Alternatively its acceptability may be dependent on the possibility of demonstrating the outcomes, probabilistically or otherwise, of using this procedure in any class of problems[4]. The latter form of justification meets the difficulty that the higher level problems not only have great uncertainty in outcome, but, to date, have rarely been subject to such formal procedures.

[1] A by no means easy problem, if these are not given in advance.

[2] Thus a revision of the initial alternatives, or of the consequence variables, may take place at any time during the analysis.

[3] The selection of any procedure generates a problem in its own right and this produces a circular process since we then have the situation in which a procedure, P, is being applied to a problem, Q, one of whose alternatives is P.

[4] A great deal of work in 'heuristic programming' is now taking place, the success of which is measured in precisely this way.

It is important to realize that the use of mathematical models, of one kind of another, for problem-solving, is subject to risk as is any other action in any other problem to one degree or another. It is not necessarily true that the returns from the use of any such analysis will offset any costs involved, and, indeed, it is surprising how often experience alone, in repetitive situations, achieves a level of performance difficult to beat by further analysis.[1]

Any problem solving procedure is 'portable' in that it will apply to all problems, being a meta partial decision rule. The theories of value and uncertainty in Chapter (2) are also said to satisfy the 'Portability Doctrine' in that one does not carry around ready made values and uncertainty measures, bearing in mind the variety and sequential connectedness of our problem areas, except in special cases where they can be shown to be basic (indecomposable) and invariant.[2] One rather applies the principles to the problems as and when they arise. Admittedly this means that we have failed to connect our problematic situations, even to a reasonable extent, but the low bound (relative to our problem complex) placed on us by the economic resources needed to carry through a connected analysis is very restrictive on the attainment of this degree of analysis.

Three different kinds of value were found to be important in problem solving, which we term 'decomposable', 'indecomposable' or 'partially decomposable', and their treatment must be different. A value is decomposable of it is induced by its consequences which have values themselves. Thus money is indecomposable if we assume it has no consequences.[3] It becomes decomposable if we consider its uses and not its intrinsic value. Thus it may be valued by valuing (choicewise) the set of all possible uses to which it can be put. Personnel satisfaction may be valued for itself, and may form an indecomposable value. On the other hand it may also be valued as a producer of system performance, measured in some other terms. It

---

[1] This statement is made, substantially, on the basis of the frequency of such assertions conveyed to me by various people over a number of years.

[2] Thus it may be useful to have a standard value function for a person's monetary values, if we can express a variety of problems in terms of indecomposable monetary consequences.

[3] Other than direct pleasure, which we omit since this is not measurable, as yet, since although we can set up an experiment in which money is specified, we cannot do so in terms of pleasure. However, we may have to change our view if Bonhert's theory (see (12)) is accepted, which is expressed in terms of basic 'units of pleasure'. The principle we are putting over will be unchanged in any case.

it likely that it is valued for both reasons, and hence such values are partially decomposable. Personnel valuations in relationship to their suitability for posts are, in principle, fully decomposable, unless one has personal tastes for certain personalities and ages.

The importance of the recognition of the three value classifications for problem solving lies in the fact that there will inevitably arise, in all problem solving procedures, the need for 'value judgements', which are equivalent to treating a decomposable value as indecomposable. Thus when formulating a problem it may be our judgement that certain alternatives can be ruled out without carrying out an investigation of these alternatives which may be feasible in principle. This, in effect, imputes a lower value to these alternatives than the ones eventually considered, and arises quite frequently in simulation studies. Value judgements are made when assessing the effects of certain actions on, say, market response.[1] The above remarks are also relevant to the value of, say, mathematical models in problem analysis. The limited evidence on the usefulness of such, in certain areas, may reduce such valuations to a purely subjective indecomposable level, depending, perhaps, on blind faith more than realistic evidence.

We thus see that one limit on the extent to which decidability can be brought into problem solving will depend on the extent to which the indecomposable elements[2] can be identified and measured. Studies in (73) and (74) illustrate these difficulties.

The value aspect mentioned in the last paragraph has an important bearing on the notion of 'objectives', a term having an essential role in practical problem solving processes. This is a word in common usage by business men in the context of problematic situations, the implication being that their actions are purposeful in a particular sense. We encounter such statements as 'it is our objective to make as much money as possible; to keep our employees as happy as possible; to give our shareholders the maximum return; to keep our costs at a minimum'. Apart from the fact that some of the objective variables are not, at present, measurable in any strict sense (see Chapter (3), Section (3)), such propositions as 'to make as much money as possible' are, in view of the remarks in Chapter (4), Section (1), empty statements in the context of problem solving. What is really meant is that 'we place a value on money, employee

---

[1] The choice implicit in this judgement is the choice of the truth values of propositions concerning market response.

[2] Or 'product' elements, as Ackoff calls them.

happiness, shareholder satisfaction, production costs'. Whether such values are decomposable or not is important, but, nevertheless, the propositions from which choice stems are of this form. Even in apparently clear cut cases, such as 'our objective is to minimize costs', this implies that an all-out effort is being made to do so irrespective of time and effort involved. Even then the question arises as to which costs are meant.[1] The same point applies to 'our objective is to produce a product to do a certain job'. Not only is there the question of specifying whether doing the job in any way, providing it is done, is sufficient, but also the question of whether the problems involved in producing the product are to be considered relevant, since there are different ways of going about getting this product. In other words many objectives are ambiguous. To set an objective allows some decidability. It allows certain products to be ruled out if they do not 'do the job'. It allows certain alternative actions to be ruled out in other problems. Thus 'our objective is to produce steel strip' rules out producing ice-cream. There is no reason why one should not set objectives which are operational so that it can be determined, in principle at least, within a finite time span, whether it has been achieved, but it has to be realized that such objectives are constraints placed on system operation which effectively have lexicographic values.

'Objectives' are not to be confused with 'objective function' as used in mathematical analysis, unless objectives are operationally verifiable. In the latter case, if e is an event contained in E, 'whether or not the objective has been achieved' can be paraphrased to 'whether or not e is in a particular set $F \subset E$'. We can then define a characteristic objective function, θ, on E, such that θ is maximized only if the objective is achieved. θ is defined as follows.

$$\theta(e) = 1 \text{ if } e \, \varepsilon \, F \qquad (1)$$

$$= 0 \text{ if } e \, \varepsilon \sim F$$

This is only of use in that it allows us to keep our problem solving within the language of 'optimization' in the context of problem solving. Let us now turn our attention to this for a short while, since this provides one of the barriers between some decision theorists and practical problem solvers, to whom the idea of 'maximization' is untenable.

---

[1] The process of minimizing costs also has economic costs.

First of all, as has been stressed in Chapter (2), 'maximization' is not an intrinsic concept in decision theory. If we allow Savage 'minimax-regret' type choice behaviour, then 'maximization' is meaningless. Maximization, and hence improvement, require the ability to put a value structure on all acts in all our problems.

Secondly there is a wide gulf between the 'maximizers' and the 'satisficers',[1] and it is essential to understand the two. The fact that a person is a 'satisficer' simply means that either he does not wish to consider the problems of whether improvement is possible, and the consequential ones of how to obtain such improvements, or he does not think the time and effort are worth the trouble. The latter interpretation can, with the appropriate values on time and effort, bring satisficing within the framework of maximization. If he simply chooses not to consider such things he is perfectly entitled to do so. This is then purely choice behaviour, and, as such, might very well form a premise in any decision analysis.[2] It is not to be assumed, however, that, because he does not search for improvements, he will not implement such if they are brought to his attention. In fact he must do so unless he has peculiar values such as those implicit in the rejection of suggestions because someone else has suggested them. An extended concept of maximization must be brought to bear on many problems which take into account the variables contingent on the acts necessary to sort out such improvements, the uncertainty in the fruitfulness of such searches and the fact that guaranteed improvements may not be possible. Discussed solely within the context of the problems which the person himself wishes or does not wish to consider, the apparent inconsistency between the 'optimization' and 'satisficing' points of view disappears.

Finally it is important to note that we ought to distinguish between a problem of the primary problem solver and that of a would be adviser. Although an adviser may wish to consider a problem, it is not to be inferred that the primary person should consider it. Remember that, in many cases, the consequences are only incidental to the choice, and are in no way guaranteed. The person can please himself, bearing in mind this point, whether he wishes to consider any problem. To him it need not exist.

[1] Simon is a satisficer.

[2] Thus a competitor might be able to decide what to do on the basis that he is not likely to retaliate to certain moves he (the competitor) makes which do not push him below his satisficing boundary.

Operational Research is often understood to be identified with 'the provision of information for better decision-making'. Ignoring, for the moment, the latter part of the sentence, let us consider the part that 'information' plays in choice behaviour in general, with reference to its provision, form and source. Since information plays a central role in our theories, we will devote the next chapter to it.

Summary

Having disposed of some of the major theoretical considerations in Decision Theory we are beginning to move towards the determination of its part in practical decision situations. In this context there are several major points to be discussed which centre on what it is we are really after; what it is we can, as a consequence of our subject matter, expect; and what it is, in practice, that our main subject matter looks like, and how it may be handled.

4.1 Improved Selection versus Improved Consequences

To the practical decisionmaker better decision-making is synonymous with improved consequences. Decision Theory is not necessarily concerned with this; it is a theory of consistent decision-making, which will, in deterministic situations, agree with the popular requirements; in general however, if the notion of preferred decision is valid, it is still an *a priori* notion as distinct from an *a posteriori* notion; in certain cases even this idea is not admissible. Two further points are that the costs of an investigation have to be weighed against the benefits, and that there is, in any case, no reason why certainty in itself is desirable; universal enlightenment is dangerous. There are, in some instances, ways of improving decision-making, viz: improved consequences, enlargement of scope of problem, the reduction of effort in resolution, the provision of relevant information which results in problem reduction. Improvement, if any, must be in relationship to the choice, the process, and the problem environment.

4.2 Physical versus Human Sciences

In order to establish the progress one may expect in the usefulness of Decision Theory, we must compare the established physical sciences with the human sciences in the context of choice behaviour. The physical sciences have achieved an exactness which is not possible in the human sciences, but they do contribute to the objective components of our investigation (e.g. production process behaviour) as distinct from the subjective aspect (the individual decision-maker). Decision Theory is concerned with the bringing together of the objective (controlled) and subjective (controlling) aspects of a problem.

One of the paradoxes we meet is that we are trying to find ways of verifying choices as being appropriate, but we have no way of verifying that our theories, on which this hinges, are appropriate. A theory is never right or wrong, and it is really the effect of errors in theories which should be pursued, but in this we have no theory.

137

In the more exact macro sciences, verification has reached an accepted standard. However, in the field of individual decision-making, the verification of, for example, probabilistic hypotheses, is not really feasible, especially if we are expected to verify that the long run behaviour is of a particular type. This has a bearing on probabilistic choice, the concept of which is undoubtedly useful, but the application of which falters on the verification side. We are very much in the hands of subjective verification.

In micro physics the stage has long been reached when probabilistic laws have been recognized as an essential part of describing behaviour, although the verification is on the basis of predictable macro behaviour.

In micro human behaviour (the individual) Heisenberg's principle is extremely relevant and this is discussed from several points of view, but the theory has yet to be developed.

Perhaps the single major difference between the physical and human sciences is the cognitive element attributed to the latter but not to the former. If we did not accept this then there would be no point in having a Decision Theory. The paradox is that we are searching for behavioural characteristics for the purpose of changing behaviour via the cognitive element, but we resolve this by stipulating that some behaviour is basic and invariant and other behaviour is inconsistent with this and may be changed.

The cognitive aspect is important in that, in a given type of problematic situation, the choice behaviour will depend on what it is that the individual perceives at the time of selection. If we could use the general language and represent all problems in terms of a basic set of consequences we may, objectively, resolve all problems without having to rely on the particular perceptions by the individual in a specific instance. We cannot do this, and hence, as we have seen in Chapter (2), develop a specific theory for the type of problem on hand, not knowing fully what the condition, c, is of the individual. This will not be the same as that perceived by the observer. Such a theory is not likely to be predictively good, and again leads back to the ideas of probabilistic choice. Particular instances of this arise when past history influences choice, a characteristic not present in the physical sciences in which the cognitive element is absent.

Two further differences between the physical sciences and human sciences (in the context of decision-making) are (a) the open or closed nature of the systems in terms of the generation of new conditions, and (b) the ability to isolate and repeat experiments for the system under consideration. Quite clearly if new conditions are being generated all the time, as is the case in business, then the application of scientific method to characterize behaviour may be ruled out. The problem of isolation and repetition creates difficulties in measuring values in sequential situations which are discussed in the context of von Neumann's theory of value.

4.3 Problems and Problem Solving Behaviour

A brief consideration of problems arising in practice indicates that they can be complex hierarchical structures, involving primary and also secondary problems. Problems of a higher order than the primary ones are seldom decidable and are often subjectively resolved by pure choice. Their appropriate consideration must reflect the costs and the benefits for the primary problem.

A specific individual is subjected to an arrival pattern of problematic situations and there arises a super problem of how to deal collectively with these. The notions of important and less important decisions needs to be investigated.

The resolution of problems in practice ranges from complete decidability to pure choice and varies throughout the levels of an organization. The problem of determining appropriate rules is quite important; we need to evaluate the extent to which completeness should exist, bearing in mind that there are advantages and disadvantages. A typical example is the problem of delegation of decision-making, which is a common concept, but for which no evaluation has been thoroughly made.

In certain infrequent types of problem we resort to guide lines to investigate them as and when they arise. The methodological guide lines of Operational Research are examined and shown to be ambiguous, and not always appropriate when the cost of the procedures are considered. It is stressed that important decisions arise in these secondary problems; in particular the choice of a mathematical model is not without its own influence on the final decision made and hence consequences achieved.

In practical problem-solving the Portability Doctrine becomes very important. Thus we do not carry around with us sets of value functions; we carry out the valuations at the time they are needed, because conditions change and hence value judgments change. Such a doctrine can give rise to incompatibilities in problem-solving however. The types of valuation which can arise (decomposable, indecomposable and partially decomposable) are considered, since these have a bearing on the degree of decidability achievable. We would like to do away with the Portability Doctrine and deal only in terms of decomposed values, but this is not possible as has already been stressed in Chapter 2.

We give consideration to some common language used in practical discussions of problems. Specifically we investigate the concepts of objective, objective function, optimization and satisficing. It is shown that there are certain misconceptions about these and that, in particular, if 'satisficing' were operationally defined, there is no conflict between satisficing and optimizing in situations in which they are meaningful (cf. Savage Minimax Regret argument in Section 1 of this chapter).

Finally we stress that a decision-maker and his adviser are different as individuals, and this has an important bearing on how it is that the adviser can be of service to the decision-maker. This is taken up more fully in the next Chapter in the context of information.

CHAPTER 5

Information for Decision [1]

In its most general form, information is simply a collection of propositions whose truth value is 0 or 1. The fact that uncertainty exists in the truth values of empirically based propositions may be represented by propositions about probabilities whose truth value is assumed to be 0 or 1. Thus the uncertainty in the truth value of $h_1(x) \equiv$ 'the demand will be x' is represented by a proposition $h_2(x) \equiv$ 'the probability that the demand will be x is $p(x)$' which is assumed to be true or false. If the truth value of h_2 is not known, then we may include propositions such as $h_3(x) \equiv$ 'the probability that $h_2(x)$ is true is $q(x, p(x))$'. As will be seen in Chapter (2), we are using probability-type chains. We need not use probabilities. We can for example, use propositions of the form $h_4 \equiv$ 'the demand will satisfy $\alpha \leqslant x \leqslant \beta$', and, although the demand is uncertain, h_4 is assumed to be true or false.

We thus see that, even in areas where uncertainty exists, information is only present in the content of propositions whose truth value is assumed to be 0 or 1, and that any proposition, whose truth value is not known to be 0 or 1, contains no information.[2] This may be a trivial observation, but it is an important one for decision purposes where premises are in the form of propositions assumed to be true.

There are basically two levels of information important to the operation of a business enterprise, which we might term 'state' and 'relation' type respectively.

The 'state'-type information simply describes the state of affairs in some context. For example, the state of the economy, the state of the company, the state of a department, the state of a piece of equipment, or even the state of a person, would come under the first heading.

[1] It is worth pointing out that many Operational Research studies are of the 'informative' type, without any attempt to prescribe actions, in the belief that information provides resolution. However, the fact that the information is valid for this purpose indirectly presupposes knowledge of criteria.

[2] Thus, if I say $h_1 : `x = a$', but do not know whether it is true or false, $h_1$ contains no information. But if I say $h_2 : `\text{prob}(x = a) = q$' and h_2 is known to be true, h_2 contains information about h_1 and its truth value.

The word 'state' need not be confined to present conditions. Thus a set of observations made over some elapsed time period, under specified conditions, would come within this heading, and gives rise, now, to a picture of past events. A forecast of future states (demand and economic conditions) would present, now, a picture of the future. Thus, although the information exists now, it may be about future or past events or observations.

The use of 'state' is very general and it may contain variables which are subject to choice or variables not subject to choice.[1] Thus the state might be 'the stock level at some time'. It might also be 'the stock level and the order quantity' at that time. One can distinguish between them if one wishes to, but it is useful for the second type of information to preserve simplicity by keeping one state concept. It will readily be seen that the example given below can be put in the 'condition → condition' format.

The relation type information specifies some connection between states. Thus 'if you take act a then consequence c will follow' specifies a causal relation between a and c. 'Economic condition c_1 will be followed by economic condition c_2' is another example, although it is not to be interpretted, necessarily, as 'c_1 causes c_2'. 'Relational' information allows us to deduce one set of propositions from another, whereas 'state' information contains no deductive element and is purely a means of identification.[2]

There is a very wide variety of information in an organization relevant to the various problems existing there, and this covers both the 'state' and 'relational' types. Thus we have information which indicates that a problem is to be solved (an action taken), information which specifies the condition of stocks and resources in general, information which specifies the condition of equipment, informational directives, and also information specifying future conditions. The relational type includes knowledge of the effect of certain actions on customers, suppliers and employees. It may include information specifying penalties and rewards for certain events. All this information impinges on the behaviour of the system, although evaluation of its worth and cost may be difficult.

Although information requires the specification of the truth value

---

[1] This is not strictly true, for, if we say '$x = a$', we may not be sure whether $x = a$ or $x = a \pm \delta$. It all depends on the level of discrimination obtainable and the consequences of such errors, or on the reliability of the information in the first place.

[2] In actual fact 'state' and 'measure' are synonymous as Chapter (3) will indicate.

of some proposition, it may not remove all ambiguity of consequences, or even directives. For example, the information that there
is a reward or penalty procedure in operation may not enable these
penalties or rewards to be predicted. The fact that a piece of equipment is now operating does not enable us to determine whether it
will be operating tomorrow. The fact that someone has requested
action in a particular context, does not necessarily enable the
admissible alternatives to be established without further contact with
the person requesting action.

One can provide state-type information which has a bearing on
any particular problem which may be only tenuously connected with
decidability. Thus, if an engineer has to choose between two pieces
of equipment, information concerning appearance may alter the
selection he may have otherwise made, but this could have had no
decidable content. If it could be established that 'appearance' was
correlated with performance, and if everything could be decomposed
in terms of monetary propositions, we add some decidability, and,
if a monetary discrimination rule exists, we could have complete
decidability.

The ability to determine the state of the system allows us to decide
what the state is and hence what action to take when certain relations
are given. Thus the production manager may decide his production
programme on the assumption that the stocks will be available; on
the other hand, it may be possible to determine what stocks are
feasible and this is tantamount to deciding the state of the system.
This then introduces a decidability aspect to the production decision.

Information, as far as decision is concerned, is simply the opposite
of ambiguity. Thus, the 'more' information the 'less' ambiguity and
vice versa, although we have to define what we mean by 'more' and
'less' in both cases.

In Chapter (2) we discussed the concept of entropy. If h_1 is a
proposition, then the uncertainty in the truth of h_1 or $\sim h_1$ is given
by $E = -p \log(p) - (1-p) \log(1-p)$, where we assume the truth of
the proposition h_2: 'the probability that h_1 is true is p'. The amount
of information in h_2 depends on what we knew before h_2 was
established. It is usually measured by the difference of the two
entropies corresponding to the states of information at the different
times.

If we use the entropy measure, then information about any
propositional truth will not always increase the information in a

related proposition. Thus if we use probabilities of probabilities and conceive of a probability distribution function, $F(p)$, over populations containing a proportion p of members with a particular property, then knowledge of p may actually decrease the information as to whether a particular member, taken at random from all populations, has this property. Thus, if we have two populations with $p = \cdot 5$ and $p = \cdot 9$ respectively, and if each population is equally likely, then the overall probability that a random selection will have the specified property is $\cdot 7$, whereas, if it were known that the member was in the first population its probability would be $\cdot 5$. We would be more uncertain, knowing this population was the relevant one, than in not knowing which population to which the member belonged. Similar remarks apply to the predictive and inverse forms of inference, where more observations may actually result in less certainty about population parameters.

It is not to be assumed that more information, measured in any specific sense, is always sought after. An answer to this problem can, as with other similar problems, only be obtained by observing a person's choice when he is faced with the alternative of choosing now or getting further information in the form of true propositions. Consider a simple example in which we have two acts, and in which the outcomes, with associated probabilities, are given as below.

		states	
		1	2
acts	1	1	100
	2	1	50

If the person could either choose from among acts 1 and 2 or find out, at a zero cost, which state were the true state, then all the theories of choice so far discussed would require that he would choose to find out the true state first. Suppose, however, there is a time lapse, T, between choice and realization of the consequences of an act. Then he could live in hope for a period T, of getting 100 or 50, whereas, if he finds out, now, what the true state is, he may be condemned to an expectation of 1 unit for this time period T. If there is a difference between the values imputed by observing choice and the feeling induced by the prospects in a choice once it has been made, then a person may not prefer perfect knowledge.

There are, of course, several forms in which information can be presented to people, depending on whether it is purely observational data, or whether it has been partially processed to produce inferences of various kinds. The relational type of information is, if empirically based, the result of some process which converts the observations into 'laws' which may be exact or inexact.[1] This is usually, but not always, done by someone different to the observer. This applies equally well to the predictive information of the state type[2], which may be done by someone in an advisory capacity. Such considerations raise some important problems.

The adviser may be influenced by personal motives in selecting the information to pass on to the person he is advising. In fact this is tautologous since it is a matter of personal choice for him, bearing in mind the consequences which may ensue from the use of this information. The extent to which he is affected will vary, but he may deliberately bias the information so that the information he passes over is the 'best' for him. We can easily construct a simple model to illustrate this.[3]

Suppose the person he is advising has an alternative set of acts, $\{k\}, k = 1, 2 \ldots m$, there are a set of possible states $\{j\}, j = 1, 2 \ldots n$, that $u(k, j)$ is the value of act k, when the state is j, to the person requiring the information. Suppose that the adviser chooses to pass over the information that state s will occur, and that the first person chooses according to the maximum value criterion. He will choose act $k(s)$ where

$$u(k(s), s) = \max_k [u(k, s)]. \tag{1}$$

Suppose the first person views his adviser's success in terms of the difference between the actual outcome and the one he could have got if the state had been forecast correctly, and imputes a corresponding penalty[4] given by the following equation, where $l(i, s)$ is the penalty if advice s is given and state i is realized.

$$l(i, s) = \phi(\max_k [u(k, i) - u(k(s), i)]) \tag{2}$$

If the adviser is minimizing his expected success loss, then he will

---

[1] See Helmer (37).

[2] e.g. the estimation of demand for a product.

[3] The following ideas stem from a paper by Nelson (52) on forecasting.

[4] Purely to represent the degree of success, but, in practice, realistic enough in its effects on the adviser's promotion prospects.

choose to advise on s which minimizes the following expression, in which $p(i)$ is the probability that state i will be realized.

$$l(s) = \sum_i p(i)l(i, s) \qquad (3)$$

Thus we have to realize that, if a second person, in the form of an adviser, provides the information, his information is not necessarily the 'best' for the person he is advising. If ϕ is linear in its argument, his choice reduces to maximizing

$$u(s) = \lambda \sum_i p(i) \, u(k(s), i). \qquad (4)$$

This maximizes the first person's expected return subject to his rule of taking s to be the actual state, and hence the interests of the two parties coincide. This need not be so if the adviser receives side payments, from other sources, dependent on the choice the first person makes. Thus, if we subtract a term $v(k(s), i)$, from the right-hand side of equation (2) the parties' interests are not identical.

The adviser is not restricted to giving a single number. He could, in fact, give the whole probability distribution, $\{p(i)\}$. This suffers from the fact, as pointed out in Chapter (2), that such probabilities only represent the adviser's feelings on the matter if they are purely subjective, and, even if they were objective, they obscure the premises on which they are based. This emphasizes the fact that they may be insufficient informational indicants. On the other hand, given enough different possible conditions, even $\{p(i)\}$ can be too complex for the first person, and the sort of approach suggested in the single parameter approach may be the best on 'simplicity' grounds, providing the value of simplicity is high. Nelson emphasizes this in his treatment of weather forecasting information for economic problems.

Savage minimax-regret criterion also plays a part when it comes to prescribing action for a decision-maker. Thus consider the following problem:

	States of Nature 1	States of Nature 2	
Alternatives 1	0	2	Outcomes
Alternatives 2	10	1	

If the adviser prescribes action 1 and the first state of nature turns out to be true, then the decision-maker is likely to say 'you have lost me 10 units by recommending alternative 1, since I could have got 10 units had you recommended alternative 2'. The gap between what he gets and could have got must reflect on the adviser. In this simple situation the answer is obvious, viz: give the decision-maker the complete picture; however in more complex problems this may not be possible without radically restricting the spectrum of alternatives considered.

This sort of problem is quite complex and can only be resolved by realizing that it presents a problem in no way different, in principle, to the ones for which the information is designed. It must depend on his problems and also on the relative worth[1] of the various forms of information provided by the adviser, and must take into account, not only the personal interests of the adviser, but also his 'expertise' or 'authoritative' position to give such informational figures. Quite obviously one has advisers because they can be expected to add some expertise to their analyses, but there is still no accepted way of measuring such expertise.[2] One could simply do this in terms of 'qualifications' and 'experience', although the relationship between these and the consequences of their advice may not be considered except in as much as general, ambiguous, propositions of the kind 'more qualifications and experience produce better assessments'. Another way would be to do this in terms of success or failure in the estimating abilities of the person as measured by comparing actual outcomes with information given in the first place. This is what Nelson does where the value of a particular forecast depends on its reliability and this reliability is taken into account when making a choice.

Consider, for example, an estimate of the outcome of some investment, as given by some second person. Let it be a single parameter α, with estimate $\hat{\alpha}$. Suppose the first person has to choose whether to invest or not, the investment requiring capital c. Bearing in mind the possible personal motivations and also subjective derivation of this estimate, it is not to be expected that the outcome will be $\hat{\alpha}$. He might reason thus. The past information, taken over a variety of situations, has been liable to error and the proportional deviation, Δ, from the actual value, has a frequency probability function $p(\Delta)$. Assuming this frequency characteristic to be a

---

[1] With respect to these problems.

[2] See Helmer (37) for a discussion of this.

permanent characteristic, he can decide what the probability function for actual outcome will be. If $q(\alpha|\hat{\alpha})$ is the probability function for the actual outcome α, he would derive

$$q(\alpha|\hat{\alpha}) = p((\hat{\alpha}-\alpha)/\alpha). \tag{5}$$

There is an assumption that $p(\Delta)$ will remain invariant. In actual fact this will change as new observations arise. We could use Baysian inference by postulating that $p(\Delta)$ were, say, Normal with unknown mean and variance, but this only adds a further difficulty. In addition, apart from the objective approach, we would wish to allow for a purely subjective assessment of $p(\Delta)$, based on what he knows of the adviser, which he feels is relevant. Thus the adviser may be optimistic or pessimistic, and he may have the company's interests in mind or just his own. At this point the only objective is to establish the principle.

Equation (5) then provides him with information on which to base his choice. If he cognitively operated according to the expected return rule, and if $e(c)$ is the expected return he could get by not investing in this investment opportunity, but using the cash, c, in some other way, then he would invest if

$$e(c) < \sum_{\alpha}\alpha q(\alpha|\hat{\alpha}). \tag{6}$$

If two advisers gave estimates $\hat{\alpha}$, $\hat{\beta}$ for different opportunities,[1] and, again, if expected returns were consciously used, the decision would be made on the basis of implementation costs, which may limit the choice to one or the other but not both, and on the two expected returns

$$\sum_{\alpha}\alpha q_{\alpha}(\alpha|\hat{\alpha}), \quad \sum_{\beta}\beta q_{\beta}(\beta|\hat{\beta}).$$

A much more complex situation arises when two people give different estimates for the same variable. Thus the Operational Research manager and the salesman may give different estimates for demand, based on different background knowledge and analysis. Going back to our first example, if c, $\hat{\alpha}$, $\hat{\alpha}'$ are the cost and two estimates respectively, how would he choose? What are the relevant premises? If we had a sequence of observations of estimates and

---

[1] Thus the mechanization manager may put forward a proposal for some computer, and the research manager for some research project, each being 'experts' in his own field, and each providing monetary measures.

outcomes of both advisers, we could construct a joint probability function, $p(\Delta_1, \Delta_2)$ for deviations from true values, and, for any α, we could get the corresponding probability function to that in equation (5), viz:

$$q(\alpha \,|\, \hat{\alpha}, \hat{\alpha}') = p((\hat{\alpha} - \alpha)/\alpha, (\hat{\alpha}' - \alpha)/\alpha). \tag{7}$$

These approaches only illustrate how premises may exist on which to hinge some decision. They may not be the 'appropriate' ones, but if decidability is required some premises will have to be found.

It is worth making a comment about the use of experience in problem situations. It has been argued that experience is what counts in successful business operations, and that, particularly in classes of problems with relatively few members, the use of probabilistic methods is not applicable. However, all experience is reducible in principle to counting the number of times certain things happened when certain conditions existed, and is hence identifiable with the frequency probability measure. The fact that a person can make 'reasonable' choices in infrequent problem classes usually depends on what he knows of the components of these problems. Thus, in insuring against special risks,[1] knowledge of human behaviour, political behaviour, and weather behaviour, are all brought, unconsciously or otherwise, to bear on the problem, although this particular problem may not have arisen before. All such behavioural knowledge reduces to the frequency probability counting procedure and anything else added to this is purely arbitrary. Viewed in the light of these remarks, Savage's theory of subjective probability is not as inadmissible as it might otherwise have been, and one can expect such probabilities to be frequency based to some extent. The advantage of the formal statistical analysis is that it does not miscount, although a difficulty lies in determining the relevant observations.

We thus see that information plays a central role in decision theory, in as much as the premises needed for decision are forms of information. However, very real problems do arise if we wish to talk in terms of 'better information', for which we have no universal answer at present. Ultimately these problems will only be resolved in terms of choice behaviour of the person for whom the information is provided. Only this way can we resolve the problems posed by the subjective

---

[1] See White (74) in which problems of insuring special projects is mentioned.

processing, and corresponding inaccuracies of information provided, and the problems of whether more information is justified, at any stage, in a problematic situation.

In as much as information is taken to be valid and unquestionable, then such will provide bases on which to hinge any reasoning in a problematic situation. In as much as doubts arise as to its validity, or inferences have to be made from it, it is purely a problem for the person being presented with it. Such problems can only be resolved, on a decision basis, by bringing in more information of a higher order.

The first five chapters have been concerned with the nature of decision, theories of choice, value and uncertainty; with decidability and its dependence on axiomatics and measure in general; with practical considerations in relationships to the difference between 'decision'- and 'consequence'-oriented research; with the problematic environment within which a person operates, and some of the problem solving behaviours and language; with the nature of human sciences in comparison with physical sciences; and, finally, with the concept of information and its relation to decision. It is now necessary to bring these together in an assessment of the pragmatic aspects of decision theory, which we shall do in the next chapter.

Summary

It is popularly held that 'better decision-making' arises through 'better information' and we examine the relevance of information to decision-making in this chapter.

Basically, information is a proposition accepted as being true or false. Two main types of information arise in decision-making, viz. 'state' and 'relational' types. The basic problem is the evaluation of the worth of information. Although the latter type contributes most to decidability, it cannot operate without the former type; the decision-maker may prefer to be well informed about existing conditions although we may not know their relevance for his decisions. Information can be ambiguous, but its basic role is the reduction of ambiguity and hence contributes decidability in the mathematical sense. The possibility, and desirability of measuring quantities of information is examined, with this purpose in mind, but we conclude that the relevance of such measures to decision-making is tenuous; in some circumstances positive quantities of information can actually increase the uncertainty factor; the entropy concept is once again examined in this light.

A fundamental problem arises as to whether a person would really prefer to know the consequences of his decisions, or remain in ignorance. By means of an example, involving delayed effects, we conclude that this is not the case.

Information can arise in relatively crude forms, involving collections of data, or in processed forms. The basic task of an adviser is to provide information, and we examine the relationship between him and the decision-maker, with the recognition that an advisor would be ill-advised to inform without considering the possible consequences of errors to him. Thus his value system will influence his advice.

In the same context the adviser has a choice of the degree of processing he carries out; highly processed information hides subjective assumptions made by him and, in the interests of the decision-maker, the appropriateness of his decisions may be unduly biassed; on the other hand, the reason for the existence of the advisory expert is that the decision-maker will not, in general, comprehend a mass of basic data. We then find ourselves accepting the subjective behaviour of one individual for the purpose of another; if the decision-maker could have comprehended the basic data he may well have chosen differently.

We examine the concept of 'expertise' and its verification. Particular attention is given to the dilemma faced by a decision-maker when two 'experts' give advice.

In management discussions great emphasis is placed on 'experience' as a determinant of appropriate decision-making ability. This is examined and it is concluded that the processing of experience to produce generalized propositions plays an important role in Decision Theory; such experience can be tapped, in its generalized form, and used to increase the decidability content of his decisions; reference to Savage's theory of subjective probability is made, in this context. We conclude that the concept of 'better decision-making' via the concept of 'better information' needs considerable re-examination; ultimately the answer to this problem can only be found within the framework of an appropriate theory of choice. Even if the information is not disputed, it may be better to suppress it; if there is some doubt about its validity, we need a theory to resolve our dilemma.

CHAPTER 6

Pragmatic Aspects of Decision Theory

Let us now consider the uses to which any theory of decision can be put, where we now define Decision Theory to be 'the study of decidability in problematic situations'. Detailed consideration of problems of implementation and verification are inherently discussed more fully in the preceding chapters and the following is a brief resumé of the more striking aspects.

As far as we are concerned our researches are oriented towards introducing decidability into selection processes. This can be achieved by providing information of the two kinds mentioned in Chapter (5) viz: 'state'-type information which contains no deductive content, and 'relational'-type which contains a deductive element. In this sense, decision simply requires information, although providing information is not automatically followed by decision if it is of the 'state' type.

Before proceeding with the manner in which decision can be introduced into problem resolution, let us reiterate four main points, viz:

(i) There can be choice without decision, and, bearing in mind the points raised in the preceding chapters, there is likely to remain a great deal of pure choice in all selection processes. (See Chapters (1) and (4)).

(ii) A distinction must be drawn between 'choice' and 'consequences'; the improvement of selection processes, wherever admissible, is not necessarily followed by improvement of consequences in uncertain situations. (See Chapter (4)).

(iii) 'Improvement' may be meaningless if certain choice criteria are allowed (see Chapter (2)). Consequently 'optimization' may be invalid (see Chapter (4)).

(iv) It is not the purpose of decision theory to remove uncertainty. It is a consequence of decision theory that uncertainty will be a co-determinant of choice.

Let us further add that decision theory is a combination of axiomatics and scientific method applied to selection processes.[1] The scientific side arises in the derivation and verification of the premises which form the basis of decision. The axiomatic side allows such premises to be used in deducing one set of propositions from another. It must be born in mind that the acceptance or rejection of premises is no less subject to choice than the physical actions to which they relate.[2]

It must also be realized that any changes brought about by decision theory are themselves subject to choice, and, consequently, to risk, in the same way as the problems to which they relate.[3]

If 'improvement' is valid in the context of any problem, or problem class, it may come about two ways. In one case the actual selection may remain invariant, but the procedure may be altered. In the other the actual selection may be altered. Both cases can arise together.

Let us begin with the use of decision theory for 'programming' selection processes.

6.1 The Use of Programmed Selection for Repetitive Problems

Programmed selection[4] applies to classes of well defined repetitive problems, and is carried out by a well defined decision rule which determines which action will be taken in any problem in the class. By 'repetition' we do not mean that the same problem is repeated. Rather it is repetitive in the sense that it belongs to a specified class characterized in a given way.

The advantages are that the process may be quicker, effort may be eliminated, boredom eliminated, and the freed time may be devoted to other matters. Simon also stresses the fact that the use of decision rules allows a certain predictability of choice, useful when trying to assess how people, or departments, will perform. Thus, if certain problem areas are delegated to a subordinate, it is important to know that he will deal with them in a certain manner.

The disadvantages will depend on the ability to ensure that the selections made are identical with the selections the person would

[1] Smith (64) mentions axiomatics, scientific method and axiology (concerned with values). However, value theory, as we have used it, is subject to scientific method.

[2] See Chapter (4).

[3] Thus, if we begin to use mathematical models to solve problems previously resolved by hunch, then we are changing the characteristics of certain selection behaviour.

[4] As discussed by Simon (63) and referred to in Chapter (4).

have made without the formal aids. Thus in setting premiums in insurance, depending on the class of problem, there is always the risk that pertinent information has not been considered, bearing in mind the large number of information conditions which could arise,[1] as against the relatively few discriminable conditions which do arise.[2]

There arises the problem of determining which rules are being used, bearing in mind the fact that 'stationary' rules may not exist, and that, as the environment changes, so does the 'rule'.[3] In fact the 'rule' is then a function of such environmental conditions. In cases where relative stability, for some reasonable time period, does exist,[4] then such rules may be derived either by verbal specification or observation of choice in specified problematic situations. If the variety of possible problems is small then there may be no difficulty. If, however, the variety is large, then we may have to resort to the axiomatic approach by means of which a relatively large number of problems can be made decidable from a few axioms.[5] Thus, in the insurance business, if the problems are monetarily oriented, and if probabilities are available, von Neumann's axioms will allow choice to be decided in a very large number of problems. In a non-probabilistic case, personnel selection may, for example, also be treated the axiomatic way.[6]

It is, of course, not essential to restrict ourselves to complete programming. Partial programming ability may also be useful. Such partial programming has been discussed in Chapter (4), and effectively introduces hierarchical (or lexicographic) choice, in which, for a given problem class either the action is decidable or, if not, is referred to a superior for his assessment. Thus, in some cases, there may be no doubt as to the appropriate action, or the risks may be small as far as taking the wrong action is concerned. In others, which belong to the same class but have different facets,[7] the superior may wish to retain personal selection.

---

[1] And hence requiring a complex decision rule.

[2] This means that such decision rules might be highly redundant.

[3] This is discussed in Chapter (4).

[4] For example, if the relevant economic conditions change smoothly and slowly.

[5] See Chapters (2) and (3).

[6] It may be established, for example, that a certain linear selection rule is effectively being used.

[7] Perhaps the delegate is likely to select a different action to the one he himself would choose because of different experience, knowledge or values.

If a delegate is going to take over some of the selection, his superior will want to know something about this choice behaviour. Chapter (2) is relevant here.

The above programming, partial or complete, is directed at repetitive type problems. These are the frequent type and are similar in that the alternative actions are all from a specific set, each pair of which is comparable. There are, however, streams of problems, between which the actions are not comparable, but these can also be partially programmed, in a sense, as was discussed in Chapter (4). We shall now discuss these.

6.2 The Use of a Problem Solving Procedure as a Partial Programme for a Heterogeneous Class of Problems

The set of all problems faced by a person forms a class in its own right, even though it is distinctly heterogeneous. It is to be expected that it would be both computationally and empirically impossible to determine a complete decision rule for this universal class of problems. However, in as much as a person may want at least an approach, bringing in some form of reason and hence decision, to solving these problems as they arise, some form of problem resolution process is required. This was discussed in Chapter (4) to some extent. The fact that it is an ambiguous process to some extent cannot be avoided. There may be different forms of process and there will remain the question of the appropriateness of such.[1] Such a procedure will be partly subject to choice and not decidable, except inasmuch as certain characteristics are accepted as universal rationality postulates.[2] There will always be difficulties of termination, for if, every time we are confronted with a problem in the application of such a procedure to solving a particular problem, we apply the same procedure to this, we never terminate.

The use of such a procedure may or may not call upon the theories outlined in Chapter (2). It all depends on the problems which are posed. We may wish to use valuations on factors which are strictly decomposable, as if they were not so.[3] On the other hand we may wish to determine the implications of these factors. Thus we may

---

[1] In fact Ackoff's procedure omits the feed back possibilities, such as re-formulation of problem.

[2] Such as the control phase (v) of Ackoff's method.

[3] See Chapter (4).

want to value delays to customers simply by observing choice in problematic situations or we may wish to assess the actual effects of such delays.[1] We may be faced with assessing uncertainty and hence with the selection of the method of assessment. The situation can be quite complex. Nonetheless even an ambiguous problem-solving procedure will provide some reasoning in selection processes, and would allow a partial programming of problems in general.

So far we have been concerned only with the programming of classes of problems, and not with the appropriateness of the programme. Let us now consider what decision theory can add to the appropriateness of any action in a given problematic context.

6.3 The Use of Theories of Choice in Deciding the Appropriateness of an Action

We are concerned here solely with the question 'in what sense is the choice of alternative a in problem Q the appropriate one?'[2]

When one thinks of 'improved' choice this is the usual sense meant. The alternatives can be single acts or policy type acts.

To be able to answer this, some meta criterion by which the assessment of the appropriateness of a can be made. The way in which this is achieved is by using certain premises to deduce certain inferences from taking act a. The inferences are identified with the act, and choice between the acts is decided by choice between the set of inferences. The decidable content lies in the exhibition that '"inference from act a" $\geqslant$ "inference from act b"'. Thus, given certain premises about environmental behaviour, and about future policy,[3] it may be inferred that the expected monetary return from act a is $e(a)$. If then it can be taken as a posit that the determinants of choice are expected returns, then this determines the appropriate act in Q. This approach has been applied to the selection of invest-

---

[1] By analyzing the correlation between delays and lost customers.

[2] The theories of choice discussed in Chapter (2) come within the programming section just discussed. We are, here, concerned with assessing the appropriateness of an act in problematic situations for which such theories of choice, at least in terms of the primary problem, are non-existent, although some theory of choice is needed for the basis of the analysis, but this is in terms of the ultimate consequences' of an act. To avoid confusion with the sort of improvement discussed in Section (5) of this chapter, it is essential to realize that the problem, including specification of alternatives, is given at the beginning.

[3] See Chapter (4) Section (2) for a discussion of the problem of future actions and their relevance to choice.

ment policies (White (72)) in which expected values were used to confirm (in some cases) the validity of the use of 'present worth' criteria for the selection of investment opportunities.

In general, the theories of choice, value and uncertainty in Chapter (2) form the basis of assessing the appropriateness of functional choices in all areas of activity, in terms of inferences about certain basic variables. To achieve any decidability, meta criteria of the kind used are essential. If we observed that choice in investment problems was not consistent[1] with 'present worth' valuations, but 'expected returns' was a valid criterion for problems expressed in terms of risky monetary returns, we would then have to produce other premises which would either decide between these two, or operate away their inconsistency. For example, we may argue that, for the investment-type problem, the final cash probabilistic approach is not valid, in which case both types of behaviour may be compatible. On the other hand, if it can be demonstrated, to the satisfaction of the investor, that the final cash probabilistic language, in this context, is a valid one, and he is prepared to admit his limited reasoning in his own selection, then his choice can be made decidable for him. It all depends on whether his own reasoning or the probabilistic reasoning is to be taken as valid.

Since any inferences which can be made depends on information being available, let us now turn to this aspect.

6.4 The Use of Information

Chapter (5) dealt with information in general and we need only stress certain points. Such information will form the basis of any reasoning, although the decision, if it exists, may only be partial. Chapter (3), Section (3), in dealing with measure, indicates how partial decision may arise. Thus, for example, information of the kind '$\alpha \leqslant x \leqslant \beta$' may allow certain alternatives to be removed. Chapter (5) discusses the subjective nature of much information and the problems of its acceptability and use in discussion. Thus the subjective nature of estimates was discussed in particular. As pointed out in Chapter (2), one cannot be certain that probabilistic measures are appropriate for any particular person. There are always problems of the appro-

---

[1] Thus choice based on the calculation of the probability of final cash accumulated in a given investment environment may not be the same as choice based on the probability of present worths using some interest factor.

priateness of such information, and this applies to all processed information.

Let us now turn to the purely mathematical aspects of decision theory.

6.5 The Use of Mathematical Models

Chapter (7) will give a fuller coverage of this aspect of decision analysis, and we shall confine ourselves to some brief remarks on the uses to which they can be put. In such models the criteria and the relevant information are taken as given, the validity of such being the general nature of our enquiry throughout the report.

Perhaps the main use of such models is that they allow a much larger class of alternatives to be investigated than is possible by the person himself. Thus, if we were considering a production scheduling problem, in which each alternative were identifiable by a ten-dimensional vector, and if there were ten levels of discrimination for each component[1] then there are 10^{10} possibilities, out of which he can only consider a few. Traditional optimization techniques, when optimization is applicable,[2] allow optimal solutions to be obtained without having to consider anything more than an extremely small proportion of possible alternatives.[3] The problem is much more severe when we are considering policy-type actions in stochastic situations, where, if there are S possible states and D possible actions, there are D^S possible steady-state policies. For cases where 'steady state' does not apply, if there are n stages in the process, then there are D^{nS} possible policies. If $D = 10$ and $S = 10^{10}$ as in the 10-dimensional, 10 discriminant levels problem above, and if $n = 10^{10}$, we get $10^{10^{10}}$ in the first case and $10^{10^{11}}$ in the second case.

In cases, such as Savage minimax regret, in which two optimization operations are involved,[4] the same applies although the work involved increases, still, however, being much much less than would be involved without the aid of a mathematical algorithm.

As has been stressed in Chapter (4), one must not assume automatically that, because 'optimal' or 'near-optimal' solutions can be

---

[1] He might, if not using a mathematical model, actually work this way.

[2] See Chapters (2), (3) and (4).

[3] Thus the simplex method of linear programming requires only a small number of points to be considered out of an enormous number of possibilities.

[4] The maximum regret is calculated and then the minimum of all the maxima is calculated.

obtained with a limited cost using mathematical analysis, whereas the coverage by the person himself is very limited, one should use such analysis even if the appropriate information and premises are available. Experience in the Dynamic Programming Research Project has shown that there is, in problems of such complexity, always a problem of the discrimination one builds into the model. Finer discrimination requires disproportionately more computational time if it is the 'states' we are discriminating, and proportional increases in computational time if it is the 'alternatives' we are discriminating.[1] Such problems are realistic and have to be resolved by comparing the advantages of finer discrimination with the consequential computational costs. Also, as has been pointed out in Chapter (4), there is always the risk that any level of analysis will produce selections not much better than those produced by the person himself. There are many cases in which mathematical programming is hard put to it to produce more than marginal gains. Thus the appropriate mathematical models are themselves subject to choice, if not decidable.

Apart from the services in 'extending' the alternatives set mentioned above, mathematical analysis has the advantage of its infallible deductive power in complex 'synthesized' systems.[2] Thus, in principle, one can represent the behaviour of a firm, in terms of component behaviours of its departments, in a mathematical way which can then be used to analyze the effects of certain actions in the firm, bearing in mind the interaction between the departments. The same can be done for complex production set ups, linking the inputs and outputs of the various component machines. Combining individual probabilistic elements produces some overall probabilistic behaviour for the connected system, and the mathematical calculus of probabilities is indispensable here. Many more examples can be given. In all of them the mathematical approach allows the systems, and their properties, to be represented[3] in such a manner as to allow overall behaviour to be deduced in terms of various actions. It aids reasoning as distinct from helping to get optimal solutions.

Essentially, any decidability in any problematic situation can only

---

[1] See Chapter (3).

[2] Those in which elements are combinations of other elements of one form or another.

[3] Ambiguously to some extent, but unambiguous in that the ambiguity is defined.

come from a mathematical treatment. We include the logic calculus in mathematics, although this is better termed 'metamathematics'. The extent to which it can be used, or ought to be used, will depend on the theories and points discussed in the preceeding chapters, particularly in Chapters (2) and (3) from which the mathematical models will stem. We might very well equate 'decision theory' with 'the theory of choice plus mathematics'.

The final chapters will now be devoted to mathematical models in general.

Summary

The preceding chapters have thrown up more problems than we have answers, but this should not preclude us from identifying senses in which better decision-making can be achieved. We thus consider such possibilities and, for our purposes, define Decision Theory to be 'the study of decidability in problematic situations'.

Before proceeding to outline the uses, we make four main points, viz:

(*a*) there is a distinction between 'choice' and 'decision';

(*b*) we are concerned with choosing 'alternatives' and not 'consequences', since we only have limited control over the latter;

(*c*) Decision Theory is concerned with 'consistency of choice' and the notion of 'improved choice' is a specialization of this;

(*d*) Decision Theory is not necessarily concerned with removing uncertainty.

The latter point is particularly important for those who think the task of Decision Theory is the removal of ignorance.

It also has to be stressed that the theories used are no less a matter of choice than the problems for which they are to be used, although we do not deal with meta problems of this kind here.

Finally we need to take care since we are trying to find ways of describing one type of behaviour to change other types of behaviour, and we must avoid overlaps. The concept of 'meta criteria' as used in Section (3) is particularly relevant here.

We now proceed to investigate the uses of Decision Theory. The main dichotomy we wish to draw is between routine programming of decisions on the basis of what actually happens and the appraisal of whether the decisions are appropriate or not.

6.1 The use of Programmed Selection for Repetitive Problems

We discuss the advantages and disadvantages of programming decision-making so that it remains the same as it was prior to the programming. The main difficulty is in describing the decision-making, and it is here that the theories of value and uncertainty are very relevant since the variables in which the specific problem area is expressed, have been ascertained. We also include partial programming, of which delegation of decision-making is a special case.

In this case we stress the requirement that the values of the delegates should be known prior to delegation.

6.2 The Use of a Problem Solving Procedure as a Partial Programme for a Heterogeneous Class of Problems

There exist a class of problems which, individually, are infrequent, and which, collectively are widely diverse. Widely diverse investment problems form part of this class. For this class we cannot programme them in the same sense as before, but can provide general (or specific) problem-solving procedures which tell us how to go about solving the problem, and allow for a fairly subjective content of the precise details of the specific investigation (e.g. selection of data, selection of model). This is a partial programme. Such programmes can, in due course, lead to programmes of the previous kind.

6.3 The Use of Theories of Choice in Deciding the Appropriateness of an Action

As we have seen in Section (1), it will be possible in general to find out how a person chooses in certain types of situation. We are here concerned with the 'appropriateness', and 'improvement' of this choice, where improvement is meaningful. We illustrate the main point with reference to capital investment where, using some theory of choice, it may happen that a 'present-worth' value approach is effectively used. However, it may be possible to express the investment decision in terms of probability distributions of final cash, after some period of time, and then to use a theory of choice based on this to select the appropriate investments. This may conflict with the existing decision-making behaviour and the need is then to invoke the latter criterion as a meta criterion (see Chapters (2) and (3)), thus leading to changes in decision-making behaviour.

6.4 The Use of Information

This has been discussed fully in Chapter (5), but we add the fact that, in general, information results in partial decision by reducing the size of the problem.

6.5 The Use of Mathematical Models

This is discussed in detail in the final chapter, and here we make two central points:

    (a) Mathematical models considerably enlarge our powers of coverage of alternatives, although one must not forget that operating such models at any degree of refinement can be costly.

    (b) Even without going so far as the prescriptive stage, mathematical models are very powerful when it comes to synthesizing and handling systems made up of interacting subsystems; their deductive powers are considerably in excess of human powers.

Using the broadened image of choice (even in the stages of scientific investigation) Decision Theory might well be identified with 'Theory of Choice plus Mathematics'.

Mathematical Models and Decision

From our point of view any model which can add any decidability content to selection processes comes within the area of mathematical models. Thus, for example, as is the case in queueing theory, any analysis which allows one to deduce the probabilities of certain events from other probabilities is allowable, even though this may not be carried any further in the analysis of the final selection. We would include simulation studies because they analyze the effect of various decision rules even though the final choice may be left to the person himself without any prior analysis of his values. The simulation procedure is simply a physical means of solving a given mathematical problem, which has first of all to be developed from the real system before simulation can be carried out.

It will not be the purpose of this report to give a detailed coverage of the mathematical models involved. These can be found in the appropriate references. The purpose is simply to ascertain the manner in which decision is introduced into selective processes by the use of mathematical models.

The essential mathematical content of our analysis lies in the manner in which the mathematical model, once derived, is processed to get the solution. We have mentioned simulation. However, the real power of mathematics lies in the ability to derive computational algorithms for characteristic models, which are as effective as possible. These algorithms effectively allow many more alternatives to be considered than would be possible by direct enumeration in the absence of a mathematical model.

Optimization is always relative to the problem specified. Thus, although there may exist alternatives which are better than the one eventually selected, this does not invalidate its optimality relative to the initial problem. If higher level optima are required then their attainment must be allowed for in the original problem. This may require that we introduce an alternative 'to search for more alternatives', and the relevant search costs.

It has already been pointed out that conventional optimization may not be applicable[1] either because of intransitivities or because of Savage 'minimax regret' type of behaviour.

Let us now look at some conventional mathematical models concerned with optimizing a function subject to constraints.

7.1 Optimization in n-Dimensional Point Space

The problem is presented as one of maximizing some function, ϕ, of control variables $x_1, x_2, \ldots x_n$ subject to constraints.[2] This is represented thus:

$$\max_{x, x_2, \ldots x_n} [\phi(x_1, x_2 \ldots x_n)], \tag{1}$$

subject to

$$[x_1, x_2, \ldots x_n] \, \varepsilon \, R_i, \quad i = 1, 2 \ldots m, \tag{2}$$

where the R_i's are admissible regions for the x_j's.

As a trivial example, problems in the linear programming domain take the following form.

$$\max_{x_1, x_2, \ldots x_n} [\Sigma \, c_j x_j], \tag{3}$$

subject to

$$\sum \alpha_{ij} x_j = b_i, \quad i = 1, 2 \ldots m, \tag{4}$$

$$x_j \geqslant 0, \quad j = 1, 2 \ldots n. \tag{5}$$

For integer programming the same format applies with a suitable definition of the R_i's.

For cases when the x_j's form a finite set of elements adequate coverage of conventional linear and non-linear programming can be found in Gass (29), Graves (35), Vajda (68), in which a variety of algorithms are discussed for characteristic forms.

There are optimization problems in which the x_j's form an infinite set, ordered by a cardinal index j. For example, the x_j's may be a sequence of different stock levels over some time period. Such problems may come under the topic of calculus of variations

[1] See Chapter (2).
[2] n may be infinite.

(Courant (21)). In cases where the classical calculus of variation fails, Pontryagin (56) provides extended analyses.

In reality, because of the discrete nature of measurement, the existence of a cardinal index is fallacious. However, it is convenient mathematically in some cases to treat a discrete index (or variable) as cardinal (continuous).[1]

In such cases, in which it is suitable to consider the x_j's as sequentially ordered, the technique of dynamic programming[2] may be useful (see Bellman (7), (8), (9) and White (75)).

There is a distinct difference between the classical optimization approach and the dynamic programming approach, and a direct connection of the latter with the theory of value as expounded in Chapter (2). To illustrate this, let us consider the problem of replacing equipment in a most economical manner, over an infinite time period. Let $\theta(t, m)$ be the discounted cost of operating a new piece of equipment, at absolute time t, for a period m.[3] Then, with $t_0 = 0$, the problem is to find a set of numbers $t_1, t_2, t_3, \ldots t_n \ldots t_\infty$ which minimizes

$$\phi(t_1, t_2, \ldots t_n \ldots) = \sum_{k=1}^{\infty} \alpha^{t_k - 1} \theta(t_{k-1}, t_k - t_{k-1}). \tag{6}$$

The conventional calculus approach expresses the optimality of a set $\{t_k^*\}$ in the set of necessary and sufficient equations. If $k \geqslant 1$, these optimality conditions are:

$$t_k \log(\alpha) \theta(t_k, t_{k+1} - t_k) + \alpha \frac{\partial}{\partial t_k} \theta(t_k, t_{k+1} - t_k)$$

$$+ \alpha^{t_{k-1} - t_{k+1}} \frac{\partial}{\partial t_k} \theta(t_{k-1}, t_k - t_{k-1}) = 0. \tag{7}$$

The dynamic programming approach defines $v(t)$ as the value (cost) of beginning with new equipment at time t and using an optimal policy[4] over the remaining duration. The value equation,

---

[1] See Chapter (3), Section (3), equation (7) which gives the optimal continuous solution to an essentially discrete problem.

[2] In the next section we discuss optimization in policy space, for which dynamic programming provides useful algorithms. However, it is only in as much as we are concerned with optimization in point space, and in as much as dynamic programming provides a radically different approach to conventional optimization methods, that we mention it here.

[3] This allows for technological change.

[4] A policy is a decision rule as a function of equipment age (which, in this case, is always zero) and absolute time t.

given below, then expresses the optimal behaviour of the system.

$$v(t) = \max_m[\theta(t, m) + \alpha^m v(t + m)] \tag{8}$$

Which approach is to be used depends on the merits of the problem and this is discussed more fully in (75). The essential point is that the optimality principles in the classical and dynamic programming approach are different. There is considerable similarity between this approach and the value theories of Koopmans and Smith, in Chapter (2).

The optimization problem, as we have formulated it, is not restricted to deterministic problems in which the outcome of any choice $[x_1 \ldots x_n]$ is known definitely and in which, therefore, the actual outcome can be made to be optimal. The same format applies to stochastic problems of the non-policy type. Thus, if $v(a, s)$ is the value of act a when the state of nature is s,[1] and if von Neumann's expected value theory applies, such problems of choosing the 'optimal' a may be formulated as in equations (1) and (2). If $p(s)$ is the probability of state s, we would have the problem

$$\text{maximize}_a \; \left[\sum_s p(s)\, v(a, s)\right] \tag{9}$$

If the valuation $v(a, s)$ is valid also under constraints,[2] then we can include constraints of the kind

$$a \, \varepsilon \, R. \tag{10}$$

For example, if $v(a, s)$ was actual monetary outcome then if $H(a) \equiv \{s : v(a, s) \geqslant \gamma\}$, (10) might take the form

$$\text{prob.} \, \{s \, \varepsilon \, H(a)\} \geqslant p_0. \tag{11}$$

Typical examples of stochastic programming, of the above type, exist in the inventory and quality control areas. If we are considering the problem of sampling batches containing a large number of items, if an *a priori* probability function $p(\xi)$ for proportion of defective items, ξ, exists, if the free variables are sample size, n, and critical acceptance/rejection number k, if the cost of a unit sample is c, if $r(\xi)$, $a(\xi)$ are the consequential costs of rejecting and accepting batches with quality ξ, and if $A(n, k, \xi)$, $R(n, k, \xi)$ are the probabilities

[1] See Chapter (2).
[2] See Chapter (2) on the point of constraints and values.

of accepting and rejecting batches of quality ξ using plan (n, k). (9) would become

$$\text{minimize}_{n,k} \left[cn + \sum_{\xi} p(\xi)(A(n, k, \xi)a(\xi) + R(n, k, \xi)r(\xi)) \right]. \quad (12)$$

Alternatively, if we wished to avoid high probabilities of accepting specific qualities ξ_1, and the cost of rejecting ξ is known, we would have the problem

$$\text{minimize}_{n,k} \left[cn + \sum_{\xi} p(\xi)R(n, k, \xi)r(\xi) \right] \quad (13)$$

subject to

$$A(n, k, \xi_1) \leqslant \alpha. \quad (14)$$

In the optimization problem, as we have formulated it, the concept of decision rule is not intrinsic to it, although the dynamic programming approach introduces it. The problem is not to find a decision rule (or policy) but simply to find an optimal 'point' in some n dimensional space (n may be infinite). Let us now turn to the problem of optimum decision rules.

7.2 Optimization in Policy Space

A policy, or decision rule, is simply a rule which determines, for any condition a given system may assume over some time period, what action is to be taken. It is a 'function' as distinct from a 'point', although it may be representable by a point. Thus if the condition is defined by a vector $s \equiv [s_1, s_2, \ldots s_m]$ we may restrict ourselves to linear decision rules given by the following equation, where $z \equiv [z_1, z_2, \ldots z_t]$ is a general alternative, and A a matrix:

$$z = As. \quad (1)$$

In this case A is the entity to be determined and can be represented by a point in mt dimensional space. If we are prepared to restrict ourselves to parametrizable policies in a given class, then the 'point' type optimization approach is valid.

For more general problems, however, providing the conditions and alternative acts form discrete sets, we can always maintain the format in Chapter (7), Section (1). We define a characteristic function $f(s, a)$ as follows:

$$f(s, a) = 1 \text{ if } a \text{ is taken when } s \text{ occurs} \quad (2)$$

165

= 0 if a is not taken when s occurs

Then, introducing the restriction,

$$\sum_a f(s, a) = 1, \qquad f(s, a) = 0 \text{ or } 1, \tag{3}$$

we can readily formulate our policy problem as one in a set of variables $x_{sa} = f(s, a)$. In fact Mann (46) does not require that $f(s, a) = 0$ or 1.[1]

The difficulty with such an approach is that if there are A possible acts and S possible states, then a policy is equivalent to a point in a set of A^S points, or to a point in S dimensional space where each component can take A values. Mann's equivalent problem is a linear programme in AS variables $\{x_{as}\}$. If the conditions are described by a three-dimensional variable with each component having ten values and if $A = 100$ we get a linear programme in 100,000 variables with 1,000 constraints.

When such a complexity exists, the usual approach is to resort to simulation, which automatically requires some parametrization, and hence restriction, of the class of policies considered. Each one is considered individually, and optimality conditions, which enable one policy to be selected from many without considering all, no longer apply. Information is simply provided about the effect of certain chosen policies on the system. If objective criteria do exist, then the number considered may be extended, within a parametrized class, by interpolation or extrapolation over the parameters.

An alternative approach is the dynamic programming approach, which in no way restricts the policies to be considered, but can be very heavy on computational time, although it remains to be seen whether Mann's approach, in the cases where it is applicable, is any less severe. A modification of Howard's algorithm (38), to allow only single changes in one component of a policy at each step, can be considered, but has yet to be tested. At present, work is going on at the Centre to reduce the computational burden in dynamic programming, by introducing heuristic mathematics, and the results are, so far, encouraging. (See Norman (54)).

The two basic stochastic situations, for which decision rules are required, are the 'stochastic known' type, in which the values of the

---

[1] Although, for his case of ergodic Markov chains, it turns out that this is so, i.e. optimal policies are pure as distinct from mixed (see Chapter (4) on probabilistic choice).

relevant probabilities are taken as given premises initially, and the 'adaptive' type, in which it is assumed that the probability distribution belongs to a given class, that the parameters determining the precise distribution are not known, but that an *a priori* probability distribution is known, describing the parameter locations. The difference between the situations is that the former distribution is not affected by future observations, but, in the latter case, the probability distribution of the parameters changes with new observations. From a programming point of view both situations are identical, but there are conflicts among many eminent people about the correct handling of the situation. Let us now consider the difficulties.

7.3 Classical Statistics, Baysian Optimization and the Optimal Use of Information

Classical statistics, in relation to the theory of significance tests, revolves around the derivation of the probability that certain events (observations) will occur when the population, from which these come, is specified probabilistically. It is identifiable with Direct Inference.[1] We do not know, in many situations, even what the parameters are, let alone the form. Thus, in analyzing the probability distribution for demand for a product, all we have is the actual demand and not the actual population. We can, as with significance tests, determine the probability of the observations made, given the population from which they come, but cannot determine the population from which they come. Unless, by some means, we are absolutely certain about the population and its parameters, we must allow for future observations to influence our beliefs about the actual stochastic nature of the system. We can if we wish, resort to the device mentioned in Chapter (7), Section (1), in which, if δ is a decision rule and F is any state of nature[2] we choose any δ which satisfies certain conditions, dependent on which F is actually true.

If we are to use the information existing in our past observations, and we are still assuming that some probability distribution is a real descriptor of system behaviour, then some form of *a priori* distribution is needed, although we still have problems of where it comes from.[3]

---

[1] See Chapter (2).

[2] In this case F is the true probability distribution.

[3] See Chapter (2).

Suppose that, now, our decision function, δ, is a function of information (observations to date) and relates to our system decisions and that information acquisition is not subject to choice.[1]

Suppose that a value function, $v(F, \delta)$, exists depending on the decision function δ and the true stochastic distribution F. Thus, for example, if the actual value, u, depends on the observations $[x_1, x_2, \ldots x_n]$ and the decisions $\delta_1(x_1), \delta_2(x_1, x_2) \ldots \delta_n(x_1 \ldots x_n)$, and if $p(x_1, x_2, \ldots x_n | F)$ is the conditional probability of $x_1, x_2, \ldots x_n$ given F, we have

$$v(F, \delta) = \sum_{x_1, x_2, \ldots x_n} p(x_1, \ldots x_n | F) u(x_1, x_2, \ldots x_n, \delta_1, \delta_2, \ldots \delta_n) \quad (1)$$

where

$$\delta_i \equiv \delta_i(x_1, x_2, \ldots x_i).$$

If an *a priori* probability distribution $q(F)$, for F, exists, the problem is represented thus as follows.

$$\text{maximize}_{\delta} \left[\sum_F q(F) v(F, \delta) \right]. \quad (2)$$

In certain cases the dynamic programming approach can usually tackle such problems. It can also handle cases where part of the problem is that of deciding when to terminate. For example, consider sequential sampling, in which jointly sufficient statistics for the problem are the number of defective items and non-defective items at any stage. Let $G_0(p)$ be the initial *a priori* distribution function for proportion of defectives, p. Let c be the unit sampling cost. Let $a(p)$, $r(p)$ be the costs of accepting and rejecting batches with proportion p of defectives. Then, if $v(m, n)$ is the expected cost, to termination, of sampling, beginning with m defectives and n non-defectives to date, and using an optimal policy, we obtain the equation:

$$v(m, n) = \min \begin{bmatrix} \int_p a(p)\, dG_{mn}(p) \\ \int_p r(p)\, dG_{mn}(p) \\ c + p(m, n)\, v(m+1, n) + (1 - p(m, n)) v(m, n+1) \end{bmatrix} \quad (3)$$

where

$$dG_{mn}(p) = p^m (1-p)^n dG_0(p) / \int_{\dot{u}} u^m (1-u)^n dG_0(u) \quad (4)$$

---

[1] Raiffa and Schlaiffer (57) and Wald (76) consider decision functions which extend to the selection of information as well, and hence allow for experimentation to be considered.

$$p(m, n) = \int_0^1 z \, dG_{mn}(z) \qquad (5)$$

The advantage of the Baysian approach here is that it allows the information available at present, and information forthcoming in future, to be considered at the time of making this choice. It enlarges the deductive content of our choice at the expense of committing us to the premises inherent in the use of an *a priori* form of analysis.

Although we have been talking in terms of sequential problems, in which many decisions may be required prior to termination, the same approach applies when we make only one set of observations prior to a terminal decision. The sampling problem in Chapter (7), Section (1) is of such a type. An adequate coverage of the Baysian optimization approach to a variety of situations can be found in (57), (76), (11).

It is not our task to resolve the 'classical statistical' versus 'Baysian optimization' approaches. Each adds some decidability to our problem, but only the latter allows the optimization concept to be applied and the dependence on future observations to be exploited. We can only get such deductive power by bringing in extra premises, in this case via value theory and the existence of *a priori* distributions.

Let us now turn to a discussion of the situations in which conventional optimality is invalid and yet decidability exists to some extent.

7.4 Non-Optimization Models

We have already seen, in Chapter (2), that if, in uncertain situations, the Savage minimax regret criterion is valid, then no valuation exists and hence optimality is meaningless. Nonetheless the ability to construct algorithms which will determine the appropriate choice without having to go through the step-by-step procedure, involving every alternative, may be developed, and, given the required need for high power discrimination, this makes an impracticable problem a practical one.

Thus if, in equation (15) of Chapter (2), we assume $\theta(a, e)$ to be a differentiable function of its arguments, and that no boundary solutions exist, the solution to the minimax-regret problem is given by

$$\frac{\partial \theta(b, e)}{\partial b} = 0,$$

169

$$\frac{\partial \theta(a, e)}{\partial a} = 0,$$

$$\frac{\partial}{\partial e}(\theta(a, e) - \theta(b, e)) = 0.$$

Such decidability, although not within the 'optimality' category, is complete, providing some process for choosing between any two alternatives, equivalent under the criterion for the specified problem, is given. Such equivalence is meant to be 'equivalence in every respect' and not simply because of an inability to discriminate further. It is conclusive equivalence. However, we can have problems in which partial orderings are allowed, which enable many alternatives to be removed, which reduce a problem Q to a smaller problem Q^1, within which the alternatives cannot be further discriminated, not because they are not, in principle, discriminable, but rather because the residual discriminating criteria are not known. We have then what is termed 'partial decidability' and 'Pareto optimality'. Given a problem Q, the problem Q^1 is the 'Pareto optimal' problem if

$$Q^1 \subseteq Q \tag{1}$$

$$(q \, \varepsilon \, Q - Q^1) \rightarrow (\exists q^1 \, \varepsilon \, Q^1) \text{ such that } q^1 > q \tag{2}$$

$$(q^{11}, q^1 \, \varepsilon \, Q^1) \rightarrow (q^{11} \not> q^1, q^1 \not> q^{11}). \tag{3}$$

The relation $>$ is only a partial ordering among $q \varepsilon Q$.

Examples of this approach can be found in Wald (76) (in which Q is a set of decision functions, and $q > q^1$ if $r(q, F) < r(q^1, F)$ for all F in the admissible set of states of nature) and Markowitz (47) (in which an alternative q is identifiable with an act a specified by a mean μ and variance σ^2, and in which $q > q^1$ if $\mu \geqslant \mu^1, \sigma^2 > (\sigma^1)^2$ or $\sigma^2 \geqslant (\sigma^1)^2, \mu > \mu^1$).[1] In Markowitz, for example, the mathematical analysis enables his so-called 'efficient sets' (Pareto optimal) to be derived by conventional mathematical programming algorithms.

Although such 'optimality' is very limited it can afford a considerable reduction in the final set of alternatives which will be presented for final selection.

Another situation in which the conventional optimization approach fails is in the area of Game Theory.[2] In the context of zero

---

[1] μ, σ are functions of q.

[2] See (22).

sum two-person games, for cases where saddle points exist, this reduces to the Wald criterion, and, with the value of an act defined as the minimum outcome for that act, a value system does exist and optimization applies. However, once we bring in unknown probability distributions over the states, the ordering of the acts becomes partial only, if one dominates the other,[1] for, if one does not dominate the other, probability distributions can be found which will reverse the orderings of the acts, and we get a minimax regret type situation in which the appropriateness of an action depends on the set of alternatives as a whole and in which no act valuations exist. The acceptance of the traditional solution depends on the reasoning that, if each person deviates from this solution, the other person will gain by the move. In general much of the theory of games centres, not on 'optimality' as we know it, but on 'equilibrium points' of the sort mentioned. Complications enter into such analyses depending on the assumptions one makes about the uncontrollable variables, e.g. passive, antagonistic, and co-operative.

Let us now turn to the use of mathematics for information provision only, as distinct from optimization.

7.5 Mathematical Models for Informational Purposes

Any mathematical model which introduces any deductive element into a selection process is admissible in Decision Theory. Thus, if, beginning from initial probability premises, one can deduce other probabilistic propositions, this introduces an element of decision into the selection of any final act. Much mathematical analysis in problematic situations is of this kind, stopping short of any attempt to prescribe actions. Thus, in quality control, one can present the person concerned with probability propositions concerning the behaviour of a specific policy, this being deduced from simpler premises such as quality of batch.[2] In inventory control, beginning with probability propositions about demand, one can deduce the probabilistic behaviour of a specific inventory policy.

Referring back to Chapter (2), given the requisite axioms, the

---

[1] If $\{\xi(e)\}$ is a probability distribution over the events, $\{e\}$ then a dominates a^1 if

$$\sum_e \theta(a, e)\xi(e) \geqslant \sum_e \theta(a^1, e)\xi(e)$$

for all $\{\xi(e)\}$.

[2] See Chapter (7), Section (2)·

ability to deduce composite values from component values is equally useful.

The variety of propositions, which are used in problem solving, and can be deduced from other propositions, is extremely wide. They do not have to be probabilistic nor complete. One set of propositions may imply any one of a set of new propositions beyond which we cannot discriminate.[1] It is on the assumption that human deduction of complex propositions is likely to error that this section plays its part.

Having said so much about mathematical models and decidability, let us now consider some problems in the use of mathematical models in problem solving. This has been stressed in Chapter (4) and the following will be a little more specific.

7.6 Problems in Mathematical Model Building

It has been stressed in Chapter (4) that the problem of whether or not to use any mathematical model, and, if so, which to use, is no less important than the problems for which they are devised. This is a point motivated by the fact that the use of any model of any form has consequences, both directly from its operation and development, and indirectly through its influence on the choice in the primary problem for which it is designed. It would certainly be useful to treat the mathematical problem from the same point of view as any other problem.

The basic problems in mathematical analysis arise out of either uncertainties in any relation or parametric values which should form part of the model, or from computational limitations. Let us consider these separately.

7.6.1 Uncertainties in Mathematical Models

In (73) a study aimed at determining an optimum district office operating policy involved the determination of the relationship between number of agents in an area and new business per year. Past information did exist, but, because of its dependence on so many other changeable factors a great deal of uncertainty existed, not only as to its form, but as to how to determine this form. For example, it was possible to conduct a controlled experiment, although this was

---

[1] See Chapter (7), Section (4).

by no means acceptable, and also involved new problems. Any attempt to resolve such uncertainties must generate new problems by adding alternative actions[1] to the old one. Such an approach can add considerable complexity to a model and a cut has to be made somewhere, which is not subject to analysis itself. At some stage, premises have to be assumed whose consequences are not known.

Apart from complications of the above kind there are simpler problems for individual parametric values, when the forms of any relation are given. Thus, if we assumed that we had a relation of the kind

$$b = \alpha n \tag{1}$$

then we might be uncertain to some extent or other about α.

Certain approaches to resolving this type of uncertainty do not accord with any of the theories of choice so far discussed in Chapter (2). Let us suppose that α is the only unknown in our model and that the outcome of act a, when α is true, is $\theta(a, \alpha)$.

A standard approach is to find, for each α in a small region of some likely α, $(\alpha = \hat{\alpha})$, the optimal a given by

$$\frac{\partial}{\partial a}\theta(a, \alpha) = 0. \tag{2}$$

This gives a policy $a(\alpha)$. If $a(\alpha)$ and $a(\alpha + \Delta\alpha)$ do not differ too much then no further analysis is used and $a(\hat{\alpha})$ is accepted. However, if $\Delta\alpha$ is small, one would expect that $a(\alpha + \Delta\alpha)$ would not differ much, providing the appropriate continuity conditions hold, and hence this is no real test for the appropriateness of $a(\hat{\alpha})$. What one ought to do, if a sensitivity analysis were valid, would be to evaluate

$$\Delta\theta = \theta(a(\hat{\alpha} + \Delta\alpha), \hat{\alpha} + \Delta\alpha) - \theta(a(\hat{\alpha}), \hat{\alpha} + \Delta\alpha). \tag{3}$$

We should not compare policies directly, but should determine the losses made by using a policy optimal for one parametric value when another value is the true one.

Since $\dfrac{\partial\theta}{\partial a}(\alpha = \hat{\alpha}) = 0$ we see that

$$\Delta\theta = \left\{\left\{\frac{1}{2}\frac{\partial a(\alpha)}{\partial \alpha}\frac{\partial^2\theta}{\partial a^2} + \frac{\partial^2\theta}{\partial a \partial\alpha}\right\}\frac{\partial a(\alpha)}{\partial \alpha}\right\}_{\substack{a = a(\hat{\alpha}) \\ \alpha = \hat{\alpha}}} (\Delta\alpha)^2 \tag{4}$$

[1] Of a different type.

Thus it is the element inside the final bracket which should be assessed in value if small errors are to be investigated. Since $\Delta\theta$ is second order in $\Delta\alpha$ one would expect that, providing $\hat{\alpha}$ were not too far out, then, in many problems, it would be 'adequate' to use crude estimates of parameters.

If either the R.H. coefficient in equation (4) is large, or $\Delta\alpha$ can be large, then α must be treated as an unknown state of nature as in Chapter (2). This will only entail finding $a(\alpha)$ in a Savage minimax regret type of situation.[1]

It is not likely that von Neumann's expected value criterion would describe choices made by mathematical analysts in situations of uncertain parameters.

At present, especially where complexities and uncertainties are high there is likely to be as much subjective assessment in model building for problem solving as in many other problematic situations for which the models may be used.

7.6.2 Computational Problems

It was demonstrated by Bremmerman (13) that there is a definite upper bound to the rate at which computations can ever, by any means, be carried out. This rate is not, by the standard of some problems, very high. For example, if each alternative were represented by a binary number with 10,000 digits it would take many thousand years to even go through every one, let alone carry out incidental calculations of consequences.

There is, quite clearly, a problem of computational compromise in problem-solving at the high-level problem area even when the model can be unambiguously formulated and, as yet, little has been done in relating computational costs to gains from further computation. Some such work is now going on at the Centre in the dynamic programming project.

The alternatives in making a model computationally feasible are either to simplify the model so that the mathematics can be handled to cater for a wide class of alternatives,[2] or to retain the 'more realistic' model but settle for a smaller coverage of alternatives,

---

[1] In, say, a Waldian type situation, the optimum act, for a given α, is only derived when the situation is trivially one of certainty.

[2] Thus we might use a cruder state description, and hence lose information, or we might not pursue the consequences of an action outside a given area.

either mathematically,[1] or by random selection. In the latter case, which is termed the 'Las Vegas' approach, alternatives are selected randomly from a specific set.[2] A curve is then plotted of the best value of the objective criterion to date against the sample number, and the process is terminated at some 'suitable' point.

At the moment no theory exists which enables the 'suitable' point to be chosen. Given the appropriate premises, the dynamic programming approach can be used to analyze this problem solving procedure as a stochastic process. Suppose that we have drawn enough samples for us to be sure that, as a consequence of our randomization process and the distribution of objective values over the alternatives which are sampled, we are prepared to accept that the values selected will follow a probability density function $g(x)$.[3] Let z be the optimal value to date, c be the sample cost, and $f(z)$ be the expected overall return, net of sampling costs, using an optimal terminating strategy. We then obtain the equation

$$f(z) = \max \left[z, \int_x f(\max[z, x]) \, g(x)dx - c \right] \qquad (1)$$

If we were to assume that x were approximately Normally distributed[4] with unknown mean, μ, and unknown variance σ^2, but that μ, σ^2 had an initial *a priori* distribution function $G(\mu, \sigma^2)$ we could obtain the adaptive[5] counterpart of equation (1), as follows:

$$f(n, m, s^2, z) = \max \left[z, \int_x f\left(n+1, \frac{nm+x}{n+1}, \frac{ns^2+(x-m)^2}{n+1}, \right. \right. \qquad (2)$$

$$\left. \left. \max(x, z)\right) dG_n(x \,|\, m, s^2) \right]$$

---

[1] Such as restricting policies to the linear type.

[2] If we are concerned with policy type alternatives, then the set may have to be parametrized and the parameters drawn randomly separately.

[3] This means that the maximum and minimum values are known but the variables which give this solution are not.

[4] In actual fact if the bounds are finite then the Normal distribution may be a poor fit since it allows for an infinite value to arise. A better distribution would be a β type distribution, although this would be more difficult to handle computationally.

[5] If the adaptive approach is not acceptable we might try the decision function approach of Wald (Chapter (7), Section (3)). Thus, if the unknown distribution of values is F, and δ is a decision function (which might be 'take a sample of n values and take the best solution out of these') we should be able to evaluate the expected gain (composed of best value obtained and the computational costs) $g(\delta, F)$. The choice of δ would then be based on the characteristic behaviour of $g(\delta, F)$. Thus we might wish to make n as small as possible while guaranteeing that the observed best value would be within 10 per cent of the true best value 90 per cent of the time, for all F in some set $\mathscr{F}$.

where m is the mean value to date, and s^2 is the variance to date, and

$$dG_n(x|m, s^2) = \frac{\int_{\mu,\sigma^2} dF(x|\mu, \sigma^2)dG(\mu, \sigma^2)}{\int_{\mu,\sigma^2} dH_n(m, s^2|\mu, \sigma^2)dG(\mu, \sigma^2)} \qquad (3)$$

Such formulations are not meant to be in any way conclusive. It is simply that, without some guiding premises which enable our models to be built and operated, the Las Vegas procedure, as a valued means of resolving the primary problems, is arbitrary. If we build mathematical models to solve primary problems, then we ought to consider the building of mathematical models to solve the problems posed by our mathematical models.

Summary

We are here concerned with the analysis once the model has been built. We stress that any model which adds decidability to our problem is relevant to our studies. We first of all concern ourselves with optimization models in the context of (*a*) *n*-dimensional point optimization, (*b*) optimization in policy space, (*c*) optimization and information value. We then consider non-optimization models in which the objective is still to prescribe action. We then consider the use of mathematical models as information processors and finally consider some problems in mathematical model building.

7.1 Optimization in *n*-Dimensional Point Space

We state the basic problem and then discuss some of the modern forms of mathematical programming developed in this area, e.g. Linear and Non-Linear Programming. It is to be stressed that even in probabilistic situations, the 'optimality concept' is meaningful, and is covered by the general area of Stochastic Programming. We discuss the ideas of 'optimality principles' which enable search to be effectively reduced.

7.2 Optimization in Policy Space

The distinction is drawn between point optimization and function optimization, and policy (or decision rule) analysis is concerned with the latter type. In general there are an astronomical number of alternative policies for even very simple situations, and this gives rise to the need for computational algorithms for this type. The advantages of Dynamic Programming over Simulation are discussed. The distinction between stochastic and adaptive situations is made, and this leads us naturally into the next section.

7.3 Classical Statistics, Baysian Optimization and the Optimal Use of Information

In Chapter (2) we discussed subjective probability and in Chapter (5) we discussed information and its relevance to decision-making. There is a controversy between classical statistics and the Baysian approach to decision-making which evolves around the concept of '*a priori* distributions' of population parameters (see Chapter (2) on Direct and Inverse Inference). The value of the Baysian approach is that it allows us to combine our *a priori* feelings about parameters with new information as it comes along. Hence it is possible to develop ways of making optimal decisions in terms of this information as it arises. In this way the value of information is given a definite interpretation.

7.4 Non-Optimization Models

The absence of optimality does not prevent us from being able to prescribe actions, once we have established the appropriate theory of choice. We discuss this type of situation in three contexts: (*a*) Savage Minimax Regret type situation, in which the intransitivity aspects rules out the optimality concept; (*b*) situations in which, because of incomparabilities, we can only use partial decision models, giving rise to Pareto Optimality, (*c*) situations of conflict, in which the Game Theory approach is used, using concepts of equilibrium points and security levels. In cases (*a*), (*c*) the non-optimality arises because choice behaviour cannot be reduced to dichotomous type situations and depends on the set of alternatives present (see Chapter (2) on Milnor's theories). In all cases, however, we can have a theory of choice with uses for decision purposes.

7.5 Mathematical Models for Informational Purposes

So far we have been concerned with the possibilities of prescribing action. However, we can contribute towards decidability simply by the derivation of appropriate processed information. Thus the ability to derive probabilities of complex events from component probabilities comes into this category, and we may not wish, in some circumstances, to pursue the implications of this for prescribing an action dependent on this. The decision-maker makes his own choice, subjectively, in the light of the information presented to him. We do not pursue the problem (see Chapter (5)) of whether he is better off with the information.

7.6 Problems in Mathematical Model Building

It is stressed that the choice of mathematical models, their degree of refinement, and the way they are used is no less a problem than the primary problem for which they are used.

The two major problems arising in the use of such models are: (*a*) the uncertainties as to their appropriateness; (*b*) the computational difficulties in using them once the model has been built.

We may, of course, not benefit from removing such uncertainties (see comment (*a*) in opening section of Chapter (6)). Also, there will always be premises which we have to accept without being able to pursue their consequences (see Chapter (4), Section (2) on criteria errors for an extreme case). One usual way of exploring the effects of assumptions is by means of sensitivity analysis, and it is seen that popular approaches to this are methodologically invalid. The sensitivity approach may conflict with von Neumann's theory of choice.

It is important, when considering the computational aspect, to realize that there is an upper limit on possible computational rates which is very low when the reality of a problem is considered. Work is needed to relate computational effort to benefits from this effort before the computational decision is made. Reference is made, in this context, to the Dynamic Programming project going on at the same time as this research. The alternatives open to us are either to simplify the model, so that no computational reductions are needed, or to reduce, heuristically, the computational effort. We may, for example, restrict the type of decision rule which we examine, or we may use the Las Vegas sampling method. We consider how the Las Vegas model, as a model for computational decisions, can be formulated.

References

1. ACKOFF, R., *Scientific Method: Optimizing Applied Research Decisions*, John Wiley and Sons, New York, 1962.
2. ACKOFF, R., *Introduction to Operations Research*, John Wiley and Sons, New York, 1957.
3. ACKOFF, R., *Progress in Operations Research*, Vol. 1, John Wiley and Sons, New York, 1961.
4. BARANKIN, E., 'Toward an Objectivist Theory of Probability,' *3rd Berkeley Symp. on Math. Stats. and Prob.*, University of California Press, 1956.
5. BAUMOL, W., 'The Neumann-Morgenstern Utility Index—An Ordinalist View,' *J. Pol. Ec.*, **59**, February 1951.
6. BAUMOL, W. (see 67).
7. BELLMAN, R., *Dynamic Programming*, Princeton University Press, New Jersey, 1957.
8. BELLMAN, R. and DREYFUS, S., *Applied Dynamic Programming*, Princeton University Press, New Jersey, 1962.
9. BELLMAN, R., *Adaptive Control Processes—A Guided Tour*, Princeton University Press, New Jersey, 1961.
10. BIRKHOFF, G. and MACLANE, S., *A Survey of Modern Algebra*, The Macmillan Company, New York, 1953.
11. BLACKWELL, D. and GIRSHICK, M., *Theory of Games and Statistical Decisions*, John Wiley and Sons, New York, 1961.
12. BONHERT, H., 'The Logical Structure of the Utility Concept,' in *Decision Processes* by Thrall, R. M., Coombs, C. H. and Davis, R. L. (eds.), John Wiley and Sons, New York, 1957.
13. BREMERMAN, A., 'Optimization Through Evaluation and Recombination', Dept. Maths. Univ. Calif. Berkeley.
14. BROWN, R., 'Measuring Uncertainty in Business Decisions', *J. Management Studies*, **1**, 2, 1964.
15. BULLINGER, C., 'The Estimation Function in Decision-making', *J. Ind. Eng.*, **13**, 1, Jan.–Feb. 1962.
16. CARNAP, R., 'Probability as a Guide in Life', *J. Phil.*, **64**, 1947.
17. CARTER, C., 'A Revised Theory of Expectations' (see 28), in *Uncertainty in Business Decisions* by Carter, C., Meredith, G. and Shackle, G. (eds.), Liverpool University Press, 1957.
18. CARTER, C., *Review of Shackle's 'Time in Economics'* (see 53).
19. CHIPMAN, J., 'The Foundations of Utility', *Econometrica*, **28**, 2, 1960.
20. CHURCHMAN, C., *Prediction and Optimal Decision*, Prentice-Hall, Englewood Cliffs, New Jersey, 1961.

21. COURANT, R., *Differential and Integral Calculus*, Vol. 2, Blackie and Sons Ltd., London, 1952.
22. DAVIDSON, D., 'Outlines of a Formal Theory of Value', *Phil. Sc.*, **22**, April 1955.
23. DEBREU, G., 'Representation of a Preference Ordering by a Numerical Function' in *Decision Processes* by Thrall, R. M., Coombs, C. H. and Davis, R. L. (eds.), John Wiley and Sons, New York, 1957.
24. DEBREU, G., 'Stochastic Choice and Cardinal Utility', *Econometrica*, **26**, 3, 1958.
25. DINEEN, J., 'The Course of Logical Positivism', *Modern Schoolman*, **34**, 1956–57.
26. DUNLOP, W., 'The Representation of Choice', *Terminological Review, Quart. Bull.* No. 3, Sept. 1951.
27. DUNLOP, W., 'Logical Formulation used in Business Decision Investigations in the Early 1930's' (see 28).
28. GALLIE, W., 'Uncertainty as a Philosophical Problem', in *Uncertainty and Business Decisions* by Meredith, Carter, *et al.* (eds.), Liverpool University Press, 1957.
29. GASS, S. I., *Linear Programming*, McGraw Hill Book Company, New York, 1958.
30. GOOD, I., 'Deciding and Decisionary Effort', *Inf. and Control*, **4**, 2 and 3, 1961.
31. GOOD, I., 'Rational Decisions', *J.R.S.S.* (B), **14**, 1, 1952.
32. GOOD, I., 'How Rational Should a Manager be?', *Management Sc.*, **8**, 4, 1962.
33. GORMAN, W., 'A Note on "A Revised Theory of Expectations" ', *Economic Journal*, **67**, 1957.
34. GOULD, C., 'Odds, Possibility and Plausibility in Shackle's Decision Theory', *Ec. J.*, **67**, 1957.
35. GRAVES, R. L. and WOLFE, P. (eds.), *Recent Advances in Mathematical Programming*, McGraw-Hill Book Company Inc., New York, 1963.
36. HAUSNER, M., 'Multidimensional Utilities', in *Decision Processes*, by Thrall, R. M., Coombs, C. H., and Davis R. L. (eds.), John Wiley and Sons, New York, 1957.
37. HELMER, O., 'On the Epistemology of Inexact Sciences', *Management Sc.*, **6**, 1, 1960.
38. HOWARD, R., *Dynamic Programming and Markov Processes*, The Technology Press and John Wiley and Sons, New York, 1960.
39. JAMES, W. (see 25).
40. KHINCHIN, A., *Mathematical Foundations of Information Theory*, Dover Publications Inc., New York, 1957.
41. KNIGHT, F., *Risk, Uncertainty and Profit*, Reprints of Economic Classics, A. M. Kelley, Bookseller, New York, 1964.
42. KOOPMANS, T., 'Utility and Impatience', *Econometrica*, **28**, 2, 1960.
43. KRELLE, W., 'A Theory on Rational Behaviour under Undercertainty', *Metroeconomica*, **XI**, fax. 1–11 (April Agosto 1959), pp. 51–63.
44. LICHTENSTEIN, S., *Bases for Preferences Among Three-Outcome Bets*, Rand Corporation, P.2909.

45. LUCE, R. and RAIFFA, H., *Games and Decisions. Introduction and Critical Survey,* John Wiley and Sons, New York, 1957.
46. MANNE, A., *Linear Programming and Sequential Decisions. Management Sc.,* **6**, 3, 1960.
47. MARKOWITZ, H., *Portfolio Selection,* Cowles Commission Monograph No. 16, John Wiley and Sons, New York, 1959.
48. MARSHAK, J., 'Rational Behaviour, Uncertain Prospects and Measurable Utility', *Econometrica,* **18**, 2, April 1950.
49. MAY, K., 'Intransitivity, Utility and the Aggregation of Preference Patterns', *Econometrica,* **22**, 1, 1954.
50. MEREDITH, G., 'Uncertainty and Business Decisions' (see 28).
51. MILNOR, J., 'Games Against Nature', in *Decision Processes* by Thrall, R. M., Coombs, C. H., and Davis, R. L. (eds.), John Wiley and Sons, New York, 1957.
52. NELSON, R., *Weather Information and Economic Decisions,* Rand Corporation Memo. R.M.-2620-NASA, August 1960.
53. NICHOLSON, M., 'A Note on Mr. Gould's Discussion of Potential Surprise', *Ec.J.,* **58**, 272, December 1958.
54. NORMAN, J. M., 'Heuristic Methods in Dynamic Programming', Ph.D. Thesis, Manchester University, 1966.
55. O'CONNOR, D., 'Uncertainty as a Philosophical Problem' (see 28).
56. PONTRYAGIN, L. S., BOLTYANSKII, V. G., GAMKRELIDZE, R. V. and MISCHCHENKO, E. G., *The Mathematical Theory of Optimal Processes,* John Wiley and Sons, New York, 1962.
57. RAIFFA, H., and SCHLAIFFER, R., *Applied Statistical Decision Theory,* Division of Research Harvard Business School, Boston, 1961.
58. REICHENBACH, H., *The Theory of Probability,* University of California Press, Berkeley, 1949.
59. ROY, A., 'The Possibility of Empirical Testing of Alternative Theories of Uncertainty' (see 28).
60. SAVAGE, L. J., *The Foundations of Statistics,* John Wiley and Sons, New York, 1954.
61. SHACKLE, G., *Decision, Order, and Time in Human Affairs,* The University Press, Cambridge, 1961.
62. SHACKLE, G., 'A Non-Additive Measure of Uncertainty', *Rev. Ec. Studies,* **17**, 1, 1950.
63. SIMON, H. A. and MARCH, J. G., *Organisations,* John Wiley and Sons, New York, 1958.
64. SMITH, N., 'Modes of Enquiry and Research Tasks for General Systems Analysis', Paper RAC-TP-72, Department of Army (American).
65. SMITH, N., 'A Calculus for Ethics: A Theory of the Structure of Value', *Behavioural Science,* **1**, April 1956.
66. TINTNER, G., 'Foundations of Probability and Statistical Inference', *J.R.S.S.* (A), **112**, 3, 1949.
67. TURVEY, R., 'Three notes on "Expectations in Economics" ', *Economica,* **16**, 64, November 1949.
68. VAJDA, S., *Mathematical Programming,* Addison-Wesley Publishing Co., Inc., Reading, Massachusetts, 1961.

181

69. VON MISES, R., 'On the Foundations of Probability and Statistics', *Ann. Maths. Stats.*, **12**, 1941.
70. VON NEUMANN, J. and MORGENSTERN, O., *The Theory of Games and Economic Behaviour*, Princeton University Press, 1953.
71. WEAVER, W., 'Probability, Rarity, Interest and Surprise', *Scientific Monthly*, **67**, 1948.
72. WHITE, D. J., 'Dynamic Programming and the Value of an Investment Opportunity', Centre for Business Research, Working Paper No. 14, 1963.
73. WHITE, D. J., 'District Office Policy', Centre for Business Research, Working Paper No. 16, 1964.
74. WHITE, D. J., 'Some Aspects of Decision-making Arising from a Brief Study in a Single Enterprise', Centre for Business Research, Working Paper No. 3, 1963.
75. WHITE, D. J., *Dynamic Programming*, Oliver and Boyd, Edinburgh, 1969.
76. WALD, A., *Statistical Decision Functions*, John Wiley and Sons, New York, 1954.
77. WILLIAMS, B., 'The Impact of Uncertainty on Economic Theory with Special Reference to the Theory of Profits' (see 28).

Index